Ranching,

endangered species,

and urbanization

in the southwest

ENVIRONMENTAL HISTORY OF THE BORDERLANDS

Thomas E. Sheridan, series editor

Ranching,

endangered species,

and urbanization

in the southwest

SPECIES OF CAPITAL

———

Nathan F. Sayre

THE UNIVERSITY OF ARIZONA PRESS TUCSON

First paperbound printing 2005
The University of Arizona Press
© 2002 The Arizona Board of Regents
All rights reserved

07   06   05   6   5   4   3   2

Library of Congress Cataloging-in-Publication Data
Sayre, Nathan Freeman.
Ranching, endangered species, and urbanization in the Southwest :
species of capital / Nathan F. Sayre.
p. cm. — (Environmental history of the borderlands)
Includes bibliographical references (p.   ) and index.
ISBN-13: 978-0-8165-2158-6 (cloth : acid-free paper)
ISBN-10: 0-8165-2158-1
ISBN-13: 978-0-8165-2552-2 (pbk. : acid-free paper)
ISBN-10: 0-8165-2552-8
1. Buenos Aires National Wildlife Refuge (Ariz.)—History.
2. Bobwhite—Arizona—History. 3. Nature conservation—Political aspects—
Arizona—History. 4. Ranching—Arizona—History.
I. Title. II. Series.
QL84.22.A6 S28 2002
333.95'416'0979177—dc21   2002000355

To Lucia

It will be ironic if one of the last forms of the separation between abstracted Man and abstracted Nature is an intellectual separation between economics and ecology. It will be a sign that we are beginning to think in some necessary ways when we can conceive of these becoming, as they ought to become, a single discipline.

—RAYMOND WILLIAMS

contents

Figures

Maps

preface

In 1984 the U.S. Fish and Wildlife Service (FWS) announced its intention to acquire the 110,000-acre Buenos Aires Ranch in the Altar Valley southwest of Tucson, Arizona, for conversion into the Buenos Aires National Wildlife Refuge. Experiments conducted in the 1970s had identified portions of the ranch as "the only habitat in the United States presently believed to be capable of supporting a viable population of masked bobwhites" (FWS 1985: 1), a federally listed endangered subspecies extirpated from the United States in the 1890s.

The proposal met with local opposition, and public debate continued for nearly a year. Only a small part of the ranch was needed for the masked bobwhite, but the owner insisted on selling it whole. The price—nearly $9 million—struck many critics as exorbitant, and area residents feared that federal ownership would undermine the local economy and school district. Ranchers in the region argued that the Buenos Aires was one of southern Arizona's finest ranches and should continue to be used for livestock production. Wildlife agencies and environmental groups countered that endangered species could not be reduced to dollars and cents and that refuge jobs and tourism would benefit the local economy. The masked bobwhite could be reestablished on the Buenos Aires, they insisted, by buying the ranch and removing the cattle. Finally, with the support of Governor Bruce Babbitt and the state's congressional delegation, proponents of the refuge prevailed. Title transferred in the summer of 1985, and livestock have been excluded ever since.

Controversy over the Buenos Aires Refuge has persisted, smoldering in the local ranching community and occasionally flaring up in the regional or national media. Despite the removal of livestock, refuge biologists have not succeeded in establishing a self-sustaining population of masked bobwhites. More than twenty-five thousand captive-bred birds have been released on the refuge, but the successful reintroduction experiments of the late 1970s have not been replicated on any consistent basis. An ambitious program of prescribed burning, intended to simulate the "natural" fire regime of the desert grasslands, has not resulted in any demonstrable change in vegetation on the Buenos Aires. Critics, most of whom are ranchers, have beseeched journalists and politicians to scrutinize the refuge more closely, characterizing it as a waste of both taxpayer money and productive land. Their claims, however, have been steeply discounted by environmentalists, newspaper editors, and the general public, who suspect the ranchers of a selfish ulterior motive: to graze their cows on the refuge.

In an atmosphere of distrust and hyperbole, the debate over the refuge has degenerated into a largely symbolic contest over whether cattle grazing in the arid and semiarid West is "good" or "bad." The complex details of land management and environmental change on the Buenos Aires, both before and after it became a refuge, have been lost in the fray. Many environmentalists feel strongly that livestock grazing on western public lands is damaging to wildlife, vegetation, and watersheds and that it should therefore cease. The Buenos Aires Refuge is, from their perspective, a model of environmental reform. Indeed, the effects of livestock exclusion and prescribed fire on the Buenos Aires should inform the debate over public lands grazing. The inverse, however, has occurred: How one answers the abstract question "Is grazing good or bad?" has come to dictate views of the Buenos Aires. In consequence, exaggerated, misleading, or mistaken claims about the refuge abound on all sides. Meanwhile, the scientific debate over the effects of grazing is far from resolved. There is no question that damage has occurred in the Altar Valley, as it has throughout the Southwest. Heavy grazing during severe droughts in the 1890s, 1920s, and 1950s caused or contributed to soil erosion, arroyo formation, vegetation change, and the local extirpation of the masked bobwhite. The present-day effects of grazing, however, are not nearly so clear, in general or in regard to endangered species, and it is far from certain that the damage of the past can be healed simply by removing livestock.

This study aims to defuse the symbolic battle by presenting the historical facts that bear on the debate over the Buenos Aires. The environmental history of the Altar Valley has not been assembled before and is relatively unknown compared to other areas of southern Arizona. Nor has the story of the masked bobwhite been familiar to more than a few people. Both accounts must be told together. How did the landscape change over a century of use for livestock production? What roles did grazing, drought, flooding, and various management practices play in these changes? How did the masked bobwhite become endangered, and how was the Buenos Aires identified for its restoration? What has the transformation from ranch to refuge meant for the land, for the masked bobwhite and other wildlife, and for the people who work or visit there? Answers to these questions, unearthed from government reports, scientific literature, ranch records, interviews, and first-hand observations, should enable all sides to initiate a more constructive dialogue about the Buenos Aires.

The history of the Buenos Aires cannot be confined to the ranch-turned-refuge alone, however. For at least the past 120 years, outside forces have decisively shaped the land by influencing the actions of those who used, owned, or managed it. Throughout the ranch period, national and even international markets—for cattle, land, and investment capital—determined profitability and influenced key management decisions. Laws and policies enacted at state and federal levels shaped everything from access to forage to the inspection, transportation, and grading of cattle and beef. An elaborate system of state and federal agricultural research and extension agencies developed scientific information and technical innovations that had profound effects on the land. The Buenos Aires is unique in certain respects, but its history has much in common with large ranches throughout the semiarid West because the processes that shaped that history operated across much larger scales. In the refuge period, bureaucratic and legal-political forces have displaced economic ones, and the state agencies involved have changed. The importance of extra-local factors, however, has not diminished.

To understand the Buenos Aires thus requires simultaneous attention to the particularities of the place and to larger-scale social, economic, political, and scientific developments that affected the land. After more than a century of intensive use, manipulation, and management, the landscape of the Buenos Aires is like a palimpsest repeatedly overwritten by the interaction of natural processes and human activities. From the largest drainage to the smallest grass plant, every feature has a social as well as an ecological story to tell. Taken together, these stories challenge both sides of the debate over "wilderness" (Cronon 1996; Worster 1997). On the one hand, the Buenos Aires is profoundly altered from its "original" ecological condition; on the other hand, it remains a place where human ambitions depend on natural processes that elude domination or control. The Buenos Aires is, in these respects, representative of most landscapes, most rivers, and arguably many oceans. Its lessons are less obvious than those conventionally derived from the industrial city or the pristine forest, but they have broader applications and deeper significance. They testify above all to the dialectical relationship between social and ecological processes. The question is how to grasp these lessons.

This study takes its methods from history and ethnography and presents its findings almost entirely as historical narrative. It is, however, informed throughout by models and concepts from geography, anthropology, social theory, and range ecology. These may be usefully, albeit briefly, reviewed here in terms of natures, states, and capitals.

THE PRODUCTION OF NATURES

The natures explored in this inquiry are of several types. First, there is the natural world of plants, animals, soils, climate, etc., studied by natural scientists. The environmental history of southern Arizona is a small but vital field, and the Buenos Aires case contributes to several of its liveliest debates: the role of fire in desert grasslands (Bahre 1985; Dobyns 1981; McPherson 1995); the formation of arroyos (Cooke and Reeves 1976; Dobyns 1981; Hastings 1959); the conversion of grasslands to desertscrub vegetation dominated by mesquite (*Prosopis* spp.; Bahre 1991; Hastings and Turner 1965); and the role of livestock grazing and range management practices in these changes (Bahre 1991). In the broader fields of range science and ecology, meanwhile, a consensus is emerging that the Clementsian model of succession toward a single "climax community" of vegetation does not work well in arid and semiarid rangelands; alternative "states and transitions" models are being developed to account for the nonlinear vegetation dynamics observed in these areas (Westoby et al. 1989; cf. Sayre 2001). As a part of this work, research is needed to reconstruct the history of particular landscapes in an attempt to identify critical thresholds: moments when human actions and natural events pushed vegetation communities into new, relatively stable states. The history of the Buenos Aires presented here offers empirical material relevant to these debates (although hardly as precise and methodical as could be hoped).

Next, there is nature as it is imagined or conceived in various social settings. Ideas of nature among scientists, government officials, ranchers, and tourists are a central theme of this study. Of particular importance is the abstract, singular idea of nature—*Nature* with a capital *N*, so to speak—which acquired its modern referent in the early days of the Industrial Revolution. "Nature, in this new sense, was . . . all that was not touched by man, spoilt by man. . . . Nature was where industry was not. . . . There came the sense of nature as a refuge, a refuge from man; a place of healing, a solace, a retreat" (Williams 1980). The Buenos Aires Refuge, in its name and its management, invokes this ideal of nature as a timeless realm capable of mitigating the stresses of urban industrial life. It is an ideal reinforced by the widely held notion that scientific knowledge about nature is somehow timeless and universal, that nature and science intrinsically stand outside of history and apart from politics. The history of the Buenos Aires challenges this idea. Indeed, ahistorical conceptions of nature are at the root of several important errors in resource management on both the ranch and the refuge.

Finally, there is nature as the interface of environment, ideas, and human activities: the nature of human and social ecology. In recent decades, critical geographers have conceptualized this in terms of the production of nature (Henderson 1999; Smith 1984), building on earlier theories of the production of space (Harvey 1982; Lefebvre 1991). Smith (1984) rightly critiques the view that nature is simply "dominated" in capitalist society; he stresses the need to theorize nature and society as a unity from the outset. I concur but urge that more attention be paid to nature as a force of production, an astonishingly complex system that generates everything from air and water to food, wildlife, vegetation, and scenery. As much as humans manipulate natural processes, we depend on them to sustain life and to produce values of various kinds (Shepard 1991). "Control" usually means that natural production has been rendered tractable and reliable, not that it has been entirely re-placed by technology. The dialectic of nature produced and nature pro-ducing is at the heart of the story of the Buenos Aires.

All of these natures are subject to change over time, and the inter-relations among them are socially and historically determined. What I examine here are the historical and contemporary *relations to nature* practiced and instantiated by people on the Buenos Aires. Some of these relations can be directly observed in the manner of ethnography. Most, however, have been mediated in ways that make direct observation diffi-cult. They must be inferred from the available information and inter-preted in terms of social and ecological processes unfolding at numerous scales simultaneously. Although organized around the Buenos Aires as a particular place, the narrative of this study shifts to larger scales now and then to accommodate this need.

STATE AGENCIES

In recent years, anthropologists, geographers, and social theorists have struggled to define the state, and some wonder in what sense it can be said to exist at all (Alonso 1994; Corrigan 1990; Corrigan and Sayer 1985; Joseph and Nugent 1994; Mitchell 1988, 1991). Historical and ethno-graphic research into "state formation" has found a dizzying, somewhat contradictory array of activities and outcomes, prompting a kind of theo-retical agnosticism (see the remarks of Corrigan and Sayer in Joseph and Nugent 1994). Geographers, meanwhile, see a need for empirical work to concretize their macrolevel theories, whose cogency rests on viewing the state in relation to capitalist production and society as a whole (Hudson 2001). "There is never a point when *the* state is finally built within a given territory and thereafter operates, so to speak, on automatic pilot." Instead, "state actions . . . should be understood as the emergent, un-intended and complex resultant of what rival 'states within the state' have done and are doing on a complex strategic terrain" (Jessop 1990: 9, emphasis in original). The present study explores these dynamics em-pirically in an attempt to bridge the gaps between and within these subdisciplines.

Numerous state agencies have played critical roles in the history of the Buenos Aires: the General Land Office, U.S. Geological Survey, U.S.

Soil Conservation Service, U.S. Department of Agriculture, Arizona State Land Department, Arizona Game and Fish Department, and more recently, U.S. Fish and Wildlife Service. Moreover, much of the documentary record of the Altar Valley consists of government surveys of land, water, geology, agriculture, vegetation, or wildlife. There is a connection here, noted in a different context by Mitchell (1988): Objectified representations—maps, reports, paperwork, censuses, signs, and so forth—play a central role in state activities. This study examines state agencies and their representational work, understood not only as discourses or forms of discipline (Knobloch 1996), but also as moments in processes of production and reproduction. If the Buenos Aires Ranch produced beef, what does the refuge produce? What are the productive processes that its employees undertake? How do these processes compare with those of state agencies in the ranch period? And how does state production relate to private commodity production?

The idea of the state has long been linked by conservative, liberal, and Marxist theorists alike to property and property relations. Generally, the normative model of property has been private property possessed by individuals (or corporations legally construed as individuals). Even legal scholars now recognize, with apparent surprise, that property has more to do with social relations than with laws or things: For something to be your property requires, ultimately, that others be "persuaded" to recognize it as such, and many conflicts over property are resolved through informal social norms and processes (Ellickson 1991; Rose 1994). But here again, private property is the implied norm. What about public property?

For as long as the Buenos Aires has been property at all, most or all of its land base has been owned by state or federal agencies. Its history has been shaped by a multitude of superimposed, sometimes conflicting forms of property: in water, forage, fixed improvements, wild and domestic animals, and labor as well as in land. One could argue that defining, enforcing, and coordinating various kinds of property has been the central challenge of Anglo-American society in the area, from the Gadsden Purchase of 1853–54 to the present. State agencies have worked hard to meet that challenge. Many of these property forms, however, do not conform easily to the private model from which their legal status derives. To call them "public property" begs the question: How do values produced by nature become property at all? Land can be surveyed, mapped, and patented; fish, game, timber, and minerals can be "reduced to possession"; forage can be converted into privately owned livestock. But what about clean water, scenic vistas, or nongame wildlife? What about endangered species? We can recognize a public interest in these types of things, and we can authorize agencies to regulate them. In so doing, the law ascribes a kind of property status to them. But can they be property if they cannot be possessed? Of more practical concern, can they be managed if the processes that sustain them are not understood or cannot be controlled? What if enforcing the public's ascribed property right in them vests an agency with a kind of monopolistic power that early conservationists sought to curtail in the private sector?

The Buenos Aires case presents both a historical look into the processes of state formation in the United States and an ethnographic examination of how (some) state agencies produce values of various kinds through the legal constructs of property. Both may shed light on "the state effect" (Mitchell 1991), or the process by which people come to believe in the state (its existence or legitimacy). Distinctions between public and private property are central to this effect, especially insofar as they naturalize the underlying concept of property itself. At least in this case, however, the state's effects are more concrete, and they cannot be understood apart from broader processes of social and economic change, specifically those related to the production and reproduction of capital.

SPECIES OF CAPITAL

The history of the Buenos Aires spans two economic booms with contrasting ways of capitalizing on land. In the late nineteenth century, the cattle boom capitalized on free grass to produce beef. The environmental ramifications of the boom, combined with natural variability in rainfall, have bedeviled range scientists and ranchers ever since. In the post–World War II period, the suburban boom has capitalized on open, relatively inexpensive land—generally former ranch land—to produce single-family homes. Enabled by surging population growth and a wide array of state activities, the suburban boom is dramatically impacting land, water, and wildlife, although its full environmental effects remain to be assessed.

Both booms illustrate the power of capital to shape landscapes, in ways both intended and inadvertent, in pursuit of further accumulation. Taken together, they confirm David Harvey's (1985a: 150) observation: "Capitalism perpetually strives . . . to create a social and physical landscape in its own image and requisite to its own needs at a particular point in time, only just as certainly to undermine, disrupt and even destroy that landscape at a later point in time. The inner contradictions of capitalism are expressed through the restless formation and re-formation of geographical landscapes." Since about 1970 a rent gap (Smith 1996) has emerged and grown in southern Arizona: The value of private land for residential development has come to exceed its value for cattle production, even in areas remote from urban centers. Some ranchers have sold out at a handsome profit, but others find their social and cultural attachments to ranching at odds with economic forces largely beyond their control. Earlier researchers characterized these ranchers as "fundamentalist" (Smith and Martin 1972); more recent scholarship indicates that they hold out for a time, then sell and relocate to more remote areas where land is cheaper and they can continue ranching (Liffmann et al. 2000). One local rancher described the predicament this way: "Ranching is not a 'job' but rather a culture, a heritage and an entire way of life. People remain in the culture long after economic sense alone would have dictated subdivision of the land, but a point can be reached when culture can not hold out against economics" (Jim Chilton, personal communication, October 1996).

Harvey's (1982) theory of the "built environment" in capitalist society is the principal lens through which this study examines landscape change on the Buenos Aires. The approach focuses on capital flows and processes of transformation: how investments fixed in the landscape (such as buildings, factories, and infrastructure of all kinds) reflect, enable, and determine the spatial and temporal dynamics of capital accumulation at specific places and times. Although conceived principally in reference to urban centers, Harvey's theory applies with equal force to capitalist agricultural landscapes, including western rangelands. I trace the investments that ranch owners made to enable and enhance livestock grazing on the Buenos Aires and to mitigate the unintended consequences of overgrazing. The creation of the Buenos Aires Refuge was made possible, directly and indirectly, by these ranch-era investments, although most of them have been removed or allowed to deteriorate under refuge ownership.

There are, however, two important dimensions of the Buenos Aires case that Harvey's theory is poorly equipped to explain. The first is the role of the state. Many of the fixed investments made in the ranch were developed, promoted, or subsidized by state agencies in the name of resource conservation. Moreover, as property of the U.S. Fish and Wildlife Service, the Buenos Aires is now formally exempt from the land market and its rent pressures, and it no longer depends on private capital for its reproduction. In many respects, refuge supporters define its value in diametric opposition to commercial, profit-driven norms. It seems that the Buenos Aires may be as much a space of the state as of capital.

Second, some account must be made of the role of people. As Henderson (1999: 81) observes, capital must circulate through individuals (both laborers and capitalists), who may or may not "cooperate." How do the abstract, disembodied imperatives of the market relate to the actions of actual people? Harvey (2000) has acknowledged that other methods are needed, including ethnographic ones, to bridge large-scale capital flows and the minds and bodies of agents. In the case at hand, it is clear that economic rationality did not unilaterally dictate management of the Buenos Aires, even when it was a ranch. Some owners routinely lost money, and they contrived ways to insulate the operation from economic constraints. Most partook of the symbolic regalia and social status associated with ranching. Indeed, agricultural economists have suggested that ranching in the late twentieth century is partly a consumptive activity; that is, the lifestyle afforded by ranch ownership compensates for poor economic returns (Smith and Martin 1972). The refuge, in turn, draws heavily on the symbolic values of rare wildlife such as the masked bobwhite. Clearly there are values other than profit and institutions other than the market at work in the Buenos Aires case.

To address these issues, I employ the broader social theory of Pierre Bourdieu (1977, 1984, 1998). Bourdieu focuses on the relations between "the objective order"—the social and natural world that people experience—and "the subjective principles of organization" through which they perceive and evaluate what they experience (1977: 164).

(This is not a rigid dualism but a dialectical relation; if abstracted from time, it appears tautological. A full exegesis of Bourdieu's theory, however, is not possible here.) "Symbolic capital is any property (any form of capital whether physical, economic, cultural or social) when it is perceived by social agents endowed with categories of perception which cause them to know it and to recognize it, to give it value" (1998: 47). Thus, for instance, the symbolic value of ranching ultimately resides in the value that people ascribe to it, which is rooted in their experiences of it: in person, in books, in movies, etc. As this example suggests, different "species of capital" (1998: 41) are distinguished by the social forms and processes that mediate them. I am principally interested in three:

1. SYMBOLIC CAPITAL. This broad category encompasses any quality or attribute deemed valuable among a group of people, along with the social structures and practices that reproduce it as valuable. Some forms of symbolic capital are embodied in people and their actions (or, in the case at hand, in certain animals); others are objectified in things or places. All are defined, consciously or otherwise, relative to larger social fields, which are mediated in various ways and may intersect, overlap, contradict, or reinforce one another. The symbolic value of the masked bobwhite, for example, derives from several sources: its rarity relative to other species of wildlife; its relation to cattle grazing, both objectively and subjectively; its identification with the state of Arizona; and its kinship with the northern bobwhite, in particular the common mating call ("bob-white") of the two subspecies. Symbolic capitals are socially constructed, but over time they are often taken for granted as natural. Reconstructing their historical genesis is, for Bourdieu, one of the central tasks of social science.

2. ECONOMIC CAPITAL. This species of capital is distinct in that it is mediated by commodity production and exchange, including fictitious commodities such as labor and land. The market confronts people as an objective reality, even though it is ultimately a function (the integral, so to speak) of subjective opinions regarding what things are worth in money terms. The mediation of commodity value in capitalist society is "quasi-objective" and abstracted from direct experience (Postone 1993): hence the need for symbolic elaboration of economic interests and power

(ranchers "taming the frontier," developers and lenders "helping people realize the American dream," etc.). Economic capital has played an enormous role in the Buenos Aires case, primarily through fixed investments, debtor-creditor relationships, and market prices for cattle and land.

3. BUREAUCRATIC OR "STATIST" CAPITAL, mediated by information and monies distributed through a hierarchy of offices. State formation, in Bourdieu's theory, is signaled by "a shift from diffuse symbolic capital, resting solely on collective recognition, to an *objectified symbolic capital,* codified, delegated and guaranteed by the state, in a word *bureaucratized*" (1998: 50–51, emphases in original). Its magic (Coronil 1997) and its danger issue from its capacity to structure both sides of the relationship described above: the objective world (through offices, agents, and activities) and the subjective mode of apprehending that world (through classification systems, schools, legal codes, etc.). Two examples are central to the Buenos Aires case: the concept of carrying capacity, which has structured the study, management, and economic mediation of ranching for nearly a century; and the category "endangered species," which is a legal construct disguised as a natural biological fact.

The principal theoretical task of this study is to examine the processes that have produced and reproduced economic, symbolic, and bureaucratic capital over time, and specifically how different species of capital are interrelated. How, for example, the mythology of ranching grew out of the cattle boom and provided symbolic capital for a particular social and economic formation, only to become detached from cattle production and appropriated by the "new" suburban economy. Also, how the masked bobwhite acquired symbolic value as "Arizona's most famous bird" and as a victim of excessive livestock grazing, a value that the U.S. Fish and Wildlife Service tries to transform into bureaucratic capital via tourism. These social processes have had profound effects on the environment of the Buenos Aires, but they cannot be understood in ecological terms alone.

NATURE AND THE CAPITALIST STATE

By and large, environmental historians have not attempted to analyze their rich empirical data in terms of the literature in critical geography, anthropology, or social theory. William Cronon's (1991) history of Chicago and the West, for example, brilliantly chronicles the ways in which natural products and landscapes were objectified and transformed to produce commodities and finance capital, linking the metropolis with banks to the east and a vast hinterland to the west. Cronon, however, declines to explore the underlying dynamics of this complex transformation. Similarly, Marc Reisner's (1986) *Cadillac Desert* presents an engrossing story of bureaucratic competition between federal water-management agencies, mediated by the U.S. Congress on one side and private firms and landowners on the other. What this might say about relationships between our government and our economy, however, is not made clear. One exception to these incomplete narratives warrants a brief discussion here.

In his book *Rivers of Empire,* Donald Worster (1985) advances the thesis that the modern American West has been fundamentally structured by efforts to control and manipulate water. Building on the work of Karl Wittfogel, Worster examines the rise of gigantic, federally funded and controlled water projects in the twentieth century and argues that there is,

> if one looks carefully, a kind of order underlying this jumbled, discordant West. . . . It is a techno-economic order imposed for the purpose of mastering a difficult environment. People here have been organized and induced to run . . . in a straight line toward maximum yield, maximum profit. This American West can best be described as a modern *hydraulic society,* which is to say, a social order based on the intensive, large-scale manipulation of water and its products in an arid setting. . . . The hydraulic society of the West . . . is increasingly a coercive, monolithic, and hierarchical system, ruled by a power elite based on the ownership of capital and expertise. (Worster 1985: 6–7, emphasis in original)

This paragraph, in particular the last sentence, has garnered Worster more severe criticism than all of his other books combined. Many of these critics (e.g., Pisani 1989; Sheridan 1995; Walton 1992) point out that power in the West is not "monolithic" but dispersed, contested, or fragmented. "Different interest groups have cultivated, manipulated, or co-opted different agencies of the government to pursue their goals. In the process, agencies have turned against one another to defend their constituents and to grab a greater share of the federal budget" (Sheridan 1995: 359–360). The critics are right on this point, but they fail to address the substance of Worster's argument. He does not mean to argue that the West is a despotism but that it is dominated by the *capitalist state*, a mode of domination with "two faces . . . , one private and the other public, depending on which way one turns the picture"; "a closely integrated system of power that include[s] *both the state and private capitalist enterprise*" (Worster 1985: 281, 284, emphasis added). Worster's point cannot be refuted by reference to political institutions alone.

Rather than a hydraulic society, perhaps the modern American West should be understood as an example of what Henri Lefebvre (1977) provocatively termed "the state mode of production." It is a social formation "oriented simultaneously towards expanded capitalist growth, the extension of urbanization, and the territorialization of social relations within state boundaries" (Brenner 1998: 470). The Buenos Aires case illustrates all three of these characteristics in empirical detail. On first glance, the transformation from ranch to refuge appears to signal a shift from private to public, from resource exploitation to preservation, from the realm of capital to the realm of the state. On closer examination, however, the oppositions that define this perception begin to melt. Another view emerges of a symbolic struggle concerned less with ecology than with the ideological legitimation of a much larger transformation.

I have not endeavored to take this history across the border into Mexico. This is, in some respects, a gross omission: Traffic in people, livestock, and other goods across the border has been, and continues to be, significant to the local economy and community. Unlike most valleys in southern Arizona, however, the Altar Valley's hydrological boundary—the watershed divide—corresponds very closely to the international boundary. From a landscape perspective, the focus on the U.S. side makes sense. The most significant economic and political forces in the valley's history have also emanated from the north. A complementary study of the Altar Valley in Mexico would be a fascinating challenge, but it was more than this project and this book could undertake.

Acknowledgments

The research for this book took place over five years, and along the way I incurred many debts. For thirteen months (July 1996–August 1997), my wife and I lived on the Buenos Aires Refuge as volunteer caretakers of the Brown Canyon Environmental Education Center ("the lodge"), hosting groups of visitors and assisting refuge staff in various ways. I am grateful to Wayne Shifflett and Thea Ulen for approving and coordinating this arrangement, respectively. During this period, I interviewed or worked alongside most of the refuge staff—my thanks to all of them, in particular Mike Goddard, Sally Gall, Bill Kuvlesky, Guy Jontz, and Rees Madsen. Thanks also to the many visitors to the lodge who allowed me to be at once host and researcher, to answer their questions and, in turn, to pose my own. During the same period, Cindy Goddard, Aaron Flesch, and Chris Benesh treated me to numerous enjoyable bird-watching trips.

From August 1997 to September 1998 we lived in Tucson, and the research focused on archives and interviews. I am particularly grateful to the King, Chilton, and Miller families for their patience and generosity in sharing their knowledge of ranching, the valley, and its history. Librarians and officials at the University of Arizona Library–Special Collections, the Arizona Historical Society, and offices of the U.S. Bureau of Land Management, U.S. Forest Service, Natural Resources Conservation Service, U.S. Fish and Wildlife Service Regional Office, University of New Mexico–Center for Southwestern Research in Albuquerque, and Pima County government were unfailingly helpful. Mary Noon Kasulaitis shared invaluable documents that she and her family have collected over the years. Barbara Tellman, Dan Robinett, and Diana Hadley guided me to important collections I might otherwise have missed. Jeff Tannler answered every question and request I put to him at the Arizona Department of Water Resources. Wayne Pruett generously shared records and recollections from his ownership of the Buenos Aires Ranch. Stephen Williams assisted in tracking down old files of the Arizona State Land Department. Jon Souder shared documents he obtained while researching the state lands once leased to the refuge. Mike Armstrong located information on agricultural tax policies. David Ellis provided illustrations from his own collection to reproduce in this book, including map 2 and figures 3, 13, and 14. My thanks to all of them and to the many other people—more than I can list here—who generously answered my questions in personal or phone interviews.

The role of science assumed greater importance as I pieced together the history of the ranch and the refuge. I was fortunate to find many scientists willing to help a nonspecialist understand the origins, methods, findings, and blind spots of their fields: Steve Russell, Bob Steidl, Steve DeStefano, Nina King, Peter Warshall, and Brian Powell in ornithology and wildlife ecology; Mitch McClaran, Guy McPherson, Dan Robinett, and Phil Ogden in range management and grassland ecology; and David Ellis, David Brown, Steve Dobrott, John Goodwin, and Jim Levy in masked bobwhite research. Work on another book (Sayre 2001) gave me a crash course in range science that has informed the final revisions of this book. I wish to thank Kris Havstad, Jim Winder, Kirk Gadzia, Jeff Herrick, George Ruyle, Bill McDonald, Courtney White, Barbara Johnson, and Dan Robinett for their generous help with that project.

Throughout the research, the Anthropology Department of the University of Chicago gave me that combination of freedom and discipline, support and criticism necessary to push through to completion. Alan Kolata, Andy Apter, Beth Povinelli, and Moishe Postone were able and generous advisors. My thanks also to Bill Sewell, Neil Brenner, Manu Goswami, Danny Noveck, and the Social Theory Workshop for countless late-night discussions, intellectual and otherwise. Anne Ch'ien helped navigate through the administrative waters with skill and humor. Tom Sheridan of the Arizona State Museum was generous enough to join the committee and guide me as I delved into Arizona's environmental history. Finally, there is no praise high enough to match Jean Comaroff's unswerving support and ruthless intellect.

The Century Fellows Program of the University of Chicago and a Graduate Research Fellowship from the National Science Foundation provided support for the training and research that made this book possible. Dissertation write-up was supported by a Weatherhead Fellowship from the School of American Research (SAR) in Santa Fe, where I enjoyed the company and criticism of Dave and Holly Edwards, Alan Goodman, Ana Celia Zentella, Roberta Haines, and Frank Salomon. I am indebted to the staff of SAR for providing an exceptionally hospitable environment for writing. More recently, time for final revisions—much more of it than I expected—has been supported by a postdoctoral fellowship with the Agricultural Research Service–Jornada Experimental Range. My thanks, too, to Patti Hartmann and Chris Szuter at the University of Arizona Press and to three anonymous reviewers of the original dissertation manuscript. I, of course, remain solely responsible for the final product.

In 1999–2000 I was fortunate to work with the Altar Valley Conservation Alliance on their watershed assessment project, funded by the Arizona Water Protection Fund. This work enabled me to put my earlier findings to good use, to look more closely at the history of the valley as a whole, and to understand in much greater detail the current ecological conditions of the watershed. Pat King, Mary Miller, Dan Robinett, Walt Meyer, John Donaldson, Pete Phelps, Karl Ronstadt, Jim Tress, and the Alliance membership made this a uniquely rewarding research experience. The future of the Altar Valley rests largely in their hands.

Finally, my warmest thanks to those closest, whose patience, interest, love, and support formed the foundation of it all. To Mom, Dad, Hutha, Harry, Gordon, Laura, Jen, and Rich—thank you. And to Lucia and Henry—words cannot suffice.

Abbreviations

AGFD	Arizona Game and Fish Department
AHS	Arizona Historical Society, Tucson
BANWR	Buenos Aires National Wildlife Refuge
BLM	U.S. Bureau of Land Management
CID	Common-Interest Development
CSWR	Center for Southwestern Research, University of New Mexico, Albuquerque
FMO	Fire Management Officer
FWS	U.S. Fish and Wildlife Service
GLO	General Land Office
NAREB	National Association of Real Estate Boards
ORP	Outdoor Recreation Planner
SACPA	Southern Arizona Cattlemen's Protective Association
SCS	U.S. Soil Conservation Service
UAAES	University of Arizona Agricultural Extension Service, Tucson
UASC	University of Arizona Library, Special Collections, Tucson
USDA	U.S. Department of Agriculture
USFS	U.S. Forest Service
USGS	U.S. Geological Survey

Introduction

When Frank Stephens traveled through the Altar Valley and into Sonora in August 1884, he left behind his larger guns to save weight in his wagon. He would later regret the decision, but streams and wells along the way were too scarce to run the risk: every available pound had to be used to carry water. Stephens was hunting, not for game but for bird specimens to send to museums and universities in the East. He was searching, in particular, for an unusual quail, the object of a scientific controversy then four years old. Some experts said it was the same as the northern bobwhite, found across the eastern half of the United States but unknown west of the Rockies; others believed it to be another bobwhite species known from much farther south in Mexico; still others maintained that it was not a bobwhite at all. Stephens encountered the bird numerous times as he traveled south up the broad alluvial plain of the valley. He shot at them repeatedly, but his small gun had too short a range, and they eluded him for three days. Finally, he killed one not far south of the border in Mexico.

Don Pedro Aguirre Jr. was also concerned about water. A prominent businessman, Aguirre had his eye on the valley's rich perennial grasses. To graze cattle would require water, and he had conceived a plan to provide it. He would construct a large earthen reservoir in the alluvial plain near the head of the valley. The precise date of completion is unclear, but surveyors' records indicate that it held water by 1886. Roughly a hundred acres in size, it came to be known as Aguirre Lake.

A year after Stephens's trip, U.S. Deputy Surveyor George Roskruge (1885–86) began the first official survey of the Altar Valley. He divided the landscape into townships, ranges, and sections; marked the corners with piles of rocks or earth; and noted the land's potential for different uses. Over and over in his journals he wrote of the valley: "Good grass. No timber." He described the alluvial plain as flat, waterless, and densely vegetated with grass; mesquite trees lined its margins and dotted the surrounding hills. With irrigation, he judged, the rich bottomland soils could produce a variety of crops. The land, he wrote, "should be subdivided."

Over the ensuing century, the actions of Stephens, Aguirre, and Roskruge ramified and intersected to constitute the story that follows. The bird Stephens killed became the type specimen for a race previously unknown to science: *Colinus virginianus ridgwayi,* commonly known as the masked bobwhite. Most of its range lay in Mexico, but it also occupied portions of the Altar Valley and the neighboring Santa Cruz Valley in the United States. Within fifteen years, the combined effects of grazing and drought would eliminate the masked bobwhite from its U.S. range. Aguirre's reservoir was the founding act of the Buenos Aires Ranch and the beginning of large scale cattle raising on the open plains of the Altar Valley. Other artificial water sources would follow—both wells and smaller reservoirs—but none would have a greater influence on the valley, directly or indirectly, than Aguirre Lake. Roskruge's survey made both ranching and the refuge possible by defining the land as property. The framework he established enabled the lands of the Altar Valley to be claimed by settlers, classified and later leased by bureaucrats, and bought and sold in the marketplace.

RANCHING AND ENVIRONMENTAL HISTORY

This is a study of the social and environmental relations between a human activity, known as ranching, and a place called the Buenos Aires. The Buenos Aires comprises today about 116,000 acres of the Altar Valley, roughly fifty miles south-southwest of Tucson, Arizona, bordered on the south by the Mexican state of Sonora. Its size and boundaries have changed over time: It was initially smaller, and from 1909 to 1926 it was part of the much larger La Osa Live Stock and Loan Company. Throughout its history it was a major ranch in the area. In behalf of the masked bobwhite, the U.S. Fish and Wildlife Service (FWS) bought the ranch in 1985 and transformed it into a national wildlife refuge. The FWS expressly rejects and aspires to reverse the environmental legacy of the ranch. By examining the social and environmental relations that have shaped the Buenos Aires through time, I hope to elucidate the environmental history of southern Arizona, the political and environmental issues surrounding ranching in the American West today, and the sociocultural values that inform wildlife management and conservation in the United States.

Given its importance as a land use in the American West and its prominence as an environmental issue, ranching has received surprisingly little attention from environmental historians. It is by now a truism that ranching predominates in the arid and semiarid regions "beyond the hundredth meridian," where rainfall is insufficient for dry farming (Stegner 1954; see also Powell 1879; Webb 1931). Likewise, the disastrous blizzards and droughts of the 1880s and 1890s are renowned (Haskett 1935; Osgood 1929; Wagoner 1952; Webb 1931; Worster 1992). The best histories of ranching, however, are nearly silent on environmental topics. Many were written before the emergence of environmental concerns (Atherton 1961; Burton 1928; Frink et al. 1956; Graham 1960; Gressley 1959, 1966; Haley 1953; Haskett 1935; Morrisey 1950; Osgood 1929; Wagoner 1952; Webb 1931). Several more recent works remain preoccupied, like so many of their precursors, with the cultural dimensions of ranching, to the neglect of environmental factors (Dary 1981; Jordan 1993; Slatta 1990). All struggle with a general lack of historical documentation of ecological conditions. The massive body of memoirs, hagiographic biographies, and cowboy tales is of little help, and even in the rare cases where good documentation on a ranch or region survives (e.g., Remley 1993), information on vegetation and range conditions tends to be slight. Moreover, unlike mining, crop agriculture, or timber cutting, livestock grazing's environmental effects are highly variable over time and are sometimes difficult to discern, making historical reconstruction problematic.

In addition to limited documentation, two chronic problems afflict the majority of scholarship on ranching. The first is that the environmental dimensions of ranching have been examined primarily by natural scientists, most of whom pay little attention to history (but see Bahre 1991; Bahre and Shelton 1996). They ask if cattle grazing is damaging to rangelands, riparian areas, or wildlife as though the question could be answered in the abstract (Belsky et al. 1999; Fleischner 1994; Fleischner et al. 1994). There is no question that tremendous damage has been done, so their answer is almost invariably "Yes." Yet history clearly shows that grazing has sometimes been devastating and other times benign. Cattle were present in Arizona, for example, for at least 175 years before discernable damage occurred in the late nineteenth century.

The second problem is a failure on the part of historians, cultural geographers, and scholars in American studies to ask precisely what "ranching" is. Historians from Walter Prescott Webb (1931) to Donald Worster (1992) have understood ranching simply as capitalist range livestock production. Anyone producing livestock on rangelands for profit rather than subsistence is engaged in ranching. It follows that—apart from Spanish and Mexican settlements in the Southwest and California—the history of ranching in the American West is the same as the history of cattle in the region. Other scholars have tacitly ratified this conception of ranching by focusing on its cultural dimensions—the costumes, tools, techniques, and lexicon of horseback cattle raising—which can be traced back to the Old World (Jordan 1993; Slatta 1990). Although no one calls these antecedents ranching, the question of where pastoralism ends and ranching begins goes unasked.[1]

THE ALTAR VALLEY

The social and environmental histories of the Altar Valley have never been essayed together. A handful of authors have described colorful personalities and memorable events on the Buenos Aires and neighboring ranches (Bourne 1968; Coolidge 1939; Duncklee 1994; Leavengood 1993; Wilbur-Cruce 1987), but they have neglected or ignored environmental questions. Ecologists and other scientists who have studied the area have either treated it ahistorically (in surveys of bird species, for instance) or enfolded it in a larger region. Some have grouped it with the grassland valleys to the east: the Santa Cruz, San Pedro, Sulphur Spring, and San Simon (Hendrickson and Minckley 1984; Humphrey 1958). Like those valleys, it is sufficiently wet to support grassland vegetation and, by extension, large numbers of livestock. A few scholars, however, have viewed the valley as part of the lower, drier region extending to the west, which Kirk Bryan (1922, 1925) termed "the Papago Country." Like that region, it lacked perennial surface water at settlement and thus saw only sparse human use. Either way, the valley has eluded sustained attention, having neither the greater human populations and documentation of the area to the east nor the special governmental status of the Tohono O'odham (formerly Papago) Reservation immediately to the west. Several influential works on vegetation change in the region have simply ignored the Altar Valley altogether (Bahre 1991; Dobyns 1981; Hastings and Turner 1965).

In short, the Altar Valley stands in a curious, in-between position relative to the region's natural history, one that has tended to keep it at the margins of scholarly study. The documentary record is fragmentary, a problem compounded by the academic division between social and ecological research. Wagoner (1952) says nothing about the valley in his history of the cattle industry in southern Arizona. Cooke and Reeves (1976: 56) describe the Altar Valley as "very poorly documented." Another researcher characterizes it as "a quiet backwater . . . given little attention in histories of southern Arizona" (Urban 1982: 2). Livestock grazing has been the dominant human use of the valley in historical times, but the few natural scientists who have attempted to bridge social and environmental issues have treated grazing as a monolithic factor, reducing it to the presence of cows and ignoring important changes in ranch management (Cooke and Reeves 1976; Hendrickson and Minckley 1984). If an adequate history of the Altar Valley is to be reconstructed, social and environmental phenomena must be treated as coconstitutive. The practices of grazing and range management must be interrogated for clues to the sequence of events that accompanied, caused, or resulted from changes in the vegetation and hydrology of the valley.

CONDITIONS BEFORE THE CATTLE BOOM

Scholars of southern Arizona's vegetation have long recognized a fundamental difficulty: Change has been so complete that no presently existing place can be taken to demonstrate "original" conditions, and memoirs, photographs, and other records are inadequate for a comprehensive reconstruction (Griffiths 1904; Hastings and Turner 1965). This is especially the case for rangelands, which "have been managed for cattle for so long that we are uncertain about their pregrazing condition" (Bahre 1991: 123). The effort to piece together a picture is scientifically necessary because one cannot measure change without a baseline. But it runs the risk of implying that some "original" static condition existed, against which all subsequent change must be evaluated. Such a reification of what was always a dynamic environment is a socially constructed myth, one that surfaces throughout the history of the Buenos Aires. For now, a brief discussion of presettlement conditions is in order.

Located in the Basin and Range physiographic province, the Altar Valley is roughly fifty miles long and fifteen miles wide, bounded on the west by the Pozo Verde, Baboquivari, and Coyote Mountains and on the east by the San Luis, Las Guijas, Cerro Colorado, and Sierrita Mountains (map 1). Its floodwaters drain to the north, joining the Santa Cruz River northwest of Tucson and then the Gila River, which bends west and slightly south toward the Colorado River. The valley floor rises from around 2,500 feet above sea level at the north end to 3,700 feet at the divide. The surrounding mountains range from 4,700 feet in the San Luis Mountains to 7,730 feet at the top of Baboquivari Peak. By convention, the north boundary of the Altar Valley is State Highway 86 (the area north of the highway is considered the Avra Valley), and the south boundary is the border with Mexico, although the watershed divide is a few miles inside the United States. (The far south end of the valley drains south to the Rio Altar and thence to the Sea of Cortez.)

The geology of the valley explains the lack of "live" or running water (Andrews 1937). Over many millennia, erosion filled the deep valley trench with rather coarse, uninterrupted alluvium. Water runs quickly off the mountains and across a "pediment" zone before sinking into the alluvium (map 1). The pediment shares the gentle slope of the adjacent alluvial area but has very shallow soils atop bedrock. It extends one to three miles from the base of the mountains and across the valley at the south. What I will call the central valley consists of the two alluvial zones: the gently sloping, dissected, gravelly mesa land (or "upper alluvial land" in map 1) extending from the mountain pediment down to the axial drainage plain; and the axial drainage plain itself ("lower alluvial land"), a nearly flat area ranging from a quarter of a mile to a mile and a half in width and comprised of finer sandy loam soils. Both alluvial zones are several hundred feet deep and uninterrupted. Arivaca Creek, the largest source of perennial water in the watershed, disappears underground upon reaching the upper alluvium, about four miles from the axial drainage, as do smaller tributaries from the surrounding mountain

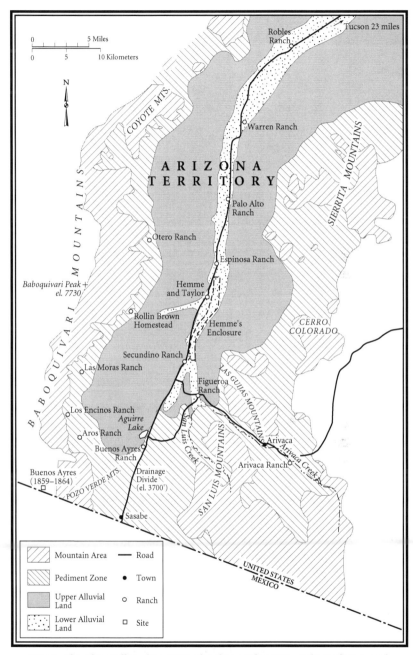

Map 1. The Altar Valley circa 1890, showing settlements, main roadways, and geological composition (Map by Don Larson, after Andrews 1937)

ranges. Wells dug in the pediment zone yield modest amounts of water from forty to one hundred feet underground; water in the central valley is abundant but deep, from two hundred to eight hundred feet below the surface.

Like the alluvium that contains it, the water underground accumulated over the centuries, building up from thousands of discrete rain events. As with other arid and semiarid regions, southwestern deserts are characterized by highly variable precipitation over both space and time. Average annual precipitation in the Altar Valley ranges from ten to twenty-four inches, depending on elevation; the Buenos Aires receives, on average, sixteen to twenty inches per year. But "average" years are so rare as to make these numbers academic. A little over half of the moisture arrives with the summer monsoon: short, intense, convectional storms in July, August, and September. The rest typically falls in December, January, and February in longer-lasting but much less intense storms from the Pacific. The fall and especially the spring months are usually very dry. Spatial variability is also high, especially in the summer: An inch or more of rain may fall over one area whereas another, less than a mile away, receives scarcely any. These patterns (or lack thereof) strongly determine the productivity and composition of vegetation. Winter rains favor some species; summer rains favor others. Net primary production of above-ground biomass can vary by a factor of ten or more between wet years and severe drought periods. Even today, rainfall remains largely unpredictable.

Because of the lack of surface water, most of what we know about the early Altar Valley is negative in character, in the sense that, while documented events were occurring nearby, little of note took place in the valley. Potsherds, manos, and metates indicate prehistoric use of the area by Hohokam and Pima Indians, but large permanent settlements have not been found.[2] Up through the Spanish and Mexican periods the valley was primarily a route between other sites, but not a major one. The route from the Arivaca Ciénega to Pozo Verde, a natural water tank south of the Baboquivari Mountains in Sonora, passed south of the valley near the present international border. Father Eusebio Francisco Kino must have traveled this route between 1687 and 1710, especially after establishing the Santa Eulalia Mission at Pozo Verde and a *visita* at Arivaca, but he never saw fit to describe the Altar Valley in writing (Urban 1982: 1). Juan Bautista de Anza, on his first trip from Tubac to California in 1774, camped for one night at Arivaca, remarking in his diary that the abundant grasses there could support more than five thousand cattle. He, too, traveled south around the Altar Valley and then west to Caborca, Sonoyta, and present-day Yuma, saying nothing about the area of concern here. The boundary survey of William Emory, published in 1857, echoes Anza's description of Arivaca; it does not even bother to give the Altar Valley a name, describing it only as "a wide and rich valley, running north and south" (1857: 121).

Two accounts from the 1860s provide our best firsthand descriptions of the Altar Valley prior to the cattle boom. Harvard mining engineer Raphael Pumpelly visited the area in 1861 as a consultant to Charles Poston. After examining Poston's Cerro Colorado Mine on the east-central margin of the valley, Pumpelly and a small party headed west toward Papago land. They crossed an area that Pumpelly dubbed "the Baboquiveri [sic] Plain," on which lie today's Palo Alto and Anvil Ranches. "This broad stretch of wild grass-land being one of the main thoroughfares of the Apaches, we were obliged to keep a good look-out all day. . . . As we entered the valley from our position on its eastern border, the broad plain lay before us[,] . . . a wide expanse of grassy steppe, and forests of mesquit [sic] and cacti. Detecting us from afar, a drove of wild horses trotted off over the grassy surface" (Pumpelly 1870: 36–37). Pumpelly also saw pronghorn, wolf, and tracks of mountain lion and bear. He remarked that the valley was little used by Apaches because of lack of water and therefore served as a buffer for the Papagos to the west.

Three years later, in 1864, popular writer J. Ross Browne traveled from Arivaca to Sasabe, following Arivaca Creek to its confluence with San Luis Wash, then turning south up the San Luis for a short distance before ascending onto Sasabe flats:

No water is found in this tract of country, but it is well wooded with mesquit [sic], and the grass is excellent. The road continues through this valley till it strikes the rise of an extensive mesa to the right, over which it continues for twelve miles. A vast plain covered with small stones and pebbles and a scanty growth of grass and cactus, bounded in the distance by rugged ranges of mountains, is all the traveller can depend upon for enjoyment during the greater part of this day's journey. It becomes oppressively monotonous after a few hours. Nothing possessed of animal life is to be seen, save at very remote intervals, and then perhaps only a lonely rabbit or a distant herd of antelope [pronghorn]. Even the smallest shrubs afford relief in this dreary wilderness of magnificent distances. . . . Descending from the mesa, as we approached the mountain range on the right, we entered a beautiful little valley, in which the grass was wonderfully luxuriant; but as usual there was no sign of water. (Browne 1871: 273–274)

Browne likely passed within a mile or two of the future site of the Buenos Aires headquarters, but he recorded no sign of settlement. He found water at Sasabe, a minor stage stop on the Mexican border (Barnes 1960). The well there was not described by William Emory and was presumably dug (in the pediment zone) between 1857 and 1864.

Pumpelly's and Browne's observations are broadly consistent with the picture ecologists have formed of the vegetation in the valley before the cattle boom: dense grass and trees in the drainages; grassy uplands nearly uninterrupted by trees or shrubs. Pumpelly passed through the Sonoran desertscrub association in the lower, hotter, and drier north end of the valley; Browne described what is now termed the Sonoran savanna grassland in the higher, moister, south end (Brown 1982). The transition occurs at roughly 3,200 feet above sea level, where palo verde, ironwood, saguaro cactus, and other desertscrub species reach the limits of their habitat. The Buenos Aires lies mostly above this line; its vegetation circa 1880 was characterized by a wide diversity of perennial bunchgrasses, annual grasses, and forbs, with a small but significant component of trees and shrubs. Most of the surrounding mountain ranges support still another community, classified as madrean evergreen woodland and characterized by evergreen oaks and junipers with an understory of grasses.

Neither Pumpelly nor Browne recorded evidence of fire, but most ecologists today agree that fire must have played a role in keeping the uplands relatively free of brush and trees (Bahre 1991; Dobyns 1981;

Humphrey 1958; McPherson 1995; Pyne 1982). Dobyns (1981) argues that the Apaches routinely set large fires for hunting, and he cites an unpublished account that the Papagos annually burned the vegetation in the neighboring Baboquivari Valley before about 1880. Lightning strikes during summer monsoons undoubtedly ignited fires as well, especially when drought parched the vegetation that had accumulated in preceding wetter years. Bahre (1985) argues that fires were "fairly frequent" in southern Arizona desert grasslands prior to intensive livestock grazing; McPherson (1995) suggests a frequency of at least once every ten years. Fire effects are highly variable, but burning generally favors grasses, which are more resilient to the disturbance than are desert grassland shrubs.

Chapter 1 explores the history of the masked bobwhite from its "discovery" and scientific classification in the 1880s to its return to the Altar Valley in the early 1970s. During this period, the masked bobwhite acquired great symbolic value among ornithologists and wildlife managers as a victim of cattle grazing, extirpated from the United States and on the brink of extinction in northern Mexico. Efforts to reintroduce the bird to the United States began in 1937 and continued into the 1960s. All were unsuccessful on the ground but critically important to making the masked bobwhite "Arizona's most famous bird." Although early advocates identified the combined effects of drought and grazing as responsible for the bird's demise, ornithologists in the 1960s blamed grazing alone. The symbolic value of the masked bobwhite helped secure its place on the first federal endangered-species list in 1969. The remaining chapters are organized chronologically.

Chapters 2 through 5 retrace the same period as chapter 1, but from the perspective of the Buenos Aires Ranch and the surrounding Altar Valley. Chapter 2 examines the conditions that Don Pedro Aguirre Jr. faced when he founded the Buenos Aires in the early 1880s. The cattle boom had pushed its way across the Great Plains and into southern Arizona, but the lack of natural water sources inhibited its spread into the Altar Valley. Aguirre's water developments made livestock production possible, and others followed his lead, especially after the drought of 1891–93 degraded rangelands elsewhere in the region. Fuelwood cutting—to supply the steam pumps that raised water from deep wells—was responsible for clearing the valley's mesquite trees, which grew along drainageways in small but significant numbers. Drought conditions from 1898 to 1904 led to severe overgrazing; cattle stripped large areas bare of vegetation before perishing themselves. Heavy rains in the winter of 1904–5 appear to have overwhelmed Aguirre's dam, initiating the arroyo that now dominates the valley's axial floodplain.

Chapter 3 examines the activities of the La Osa Live Stock and Loan Company, which owned the Buenos Aires (along with much of the rest of the valley) from 1909 to 1926. La Osa was primarily a stocker operation that imported cattle from Mexico for fattening and sale. Its managers struggled to balance the constant needs of their investors and their cattle in the face of two highly unpredictable forces: markets and weather. To try to dampen these risks, the company invested heavily in dams and further water improvements, and it worked to secure its use of the range through selection of valley lands by the Arizona State Land Department. Meanwhile, annual grasses colonized areas denuded during the boom, largely replacing perennials, and the arroyo grew rapidly. The carrying capacity of the ranch, inferred from stocking rates, gradually declined. Elsewhere in the region, government researchers began to develop scientific principles and guidelines for livestock grazing in Arizona's desert grasslands, but it does not appear that this knowledge influenced the management of La Osa.

Chapter 4 traces the Buenos Aires through the ownership of Fred Gill and Sons, from 1926 to 1959. The Gills invested heavily in fencing and further water developments, as well as in erosion control measures that were promoted by government agencies; after the depression, the U.S. Soil Conservation Service (SCS) shared the cost of most improvements. By the last half of the 1930s, perennial grasses had regained their dominance of the range. In 1948 the Gills brought in Hereford cows and switched from fattening steers to breeding calves. They ran the Buenos Aires as four ranches in one, each herd having its own cowboys, pastures, and headquarters. Late in the Gills' ownership, mesquite trees came to dominate the former grassland mesas of the ranch, and runoff rates were high. The two processes interacted in a mutually reinforcing cycle with management practices. Severe drought in the 1950s exacerbated these trends and caused the Gills to reduce the size of their herds.

Chapter 5 examines the period from 1959 to 1972. It was a time of relative inactivity on the Buenos Aires, but the larger context of ranching was rapidly shifting toward real estate speculation and development. Conditions on the ranch appear to have reached a low point in terms of stocking rates, with mesquite trees dominating the range and forage production continuing to decline. Meanwhile, suburban growth around Tucson sparked conflicts between small, undercapitalized cattle raisers and newly arrived home owners. Cattle drinking from swimming pools and feeding on expensive landscaping prompted petition drives and changes in legislation to compel their owners to fence previously open rangelands where houses were sprouting. Agricultural extension agencies began to produce pamphlets and information aimed at assisting in the creation of lawns and playgrounds for the new residents.

Chapter 6 examines the Buenos Aires under the ownership of the Victorio Company from 1972 to 1983, when masked bobwhites were re-introduced to the Altar Valley. A diversified venture capital firm, Victorio utilized ranch properties both as speculative investments and for cattle production, capitalizing on favorable tax policies and rapidly appreciating real estate values. The Buenos Aires benefited from this arrangement, as Victorio plowed profits from other activities into an expensive program of range restoration. In addition to reviving all previous owners' improvements, the company employed bulldozers to remove mesquite trees from some sixty thousand acres, seeding the treated areas with a mixture of native and non-native grasses. Concurrently, scientists from the FWS and the Arizona Game and Fish Department (AGFD) were refining their understanding of masked bobwhite habitat and perfecting techniques for releasing captive-bred birds to the wild. These two equally ambitious efforts—to restore grasses and masked bobwhites—came together in the mid-1970s, when the scientists initiated releases in bulldozed areas of the Buenos Aires that were being rested from livestock to allow grasses to establish. For three years in the late 1970s, the experiments worked: released masked bobwhites survived and bred successfully in the wild.

The effects of the range restoration efforts were not, however, figured into the analysis of the experiments. The findings were further distorted in the political process of creating the refuge in the early 1980s.

Chapters 7 and 8 focus on the Buenos Aires National Wildlife Refuge (BANWR) from its creation in 1985 to the present. Chapter 7 begins with a review of early planning efforts for the refuge and the events leading up to a major land swap, which gave the FWS ownership of the State Trust lands previously leased to the ranch. Documents strongly suggest that the FWS saw tourism—"public use," as it is called bureaucratically—as a major objective for the refuge from the beginning. This ambition focused initially on Aguirre Lake, where reliable water attracted a wide variety of birds. Livestock exclusion decreased runoff rates, however, causing the lake to cease to fill and forcing management to shift its public use plans to other areas. I also examine refuge management of the masked bobwhite and the use of prescribed fire to manipulate habitat. The refuge's activities strongly resemble those of the ranch, except that the FWS constructs "nature" differently. Anything within its power to control or manipulate is manipulated, whereas anything beyond its means is valorized as natural. By removing cattle and intentionally burning the vegetation (fire being a "natural process"), the FWS aspires to restore the Buenos Aires to its "original" condition of Sonoran savanna grassland. Because no one knows in detail what the original condition was, the masked bobwhite is both a beneficiary of this program and an indicator of its progress: In theory, when the habitat is restored, the birds will survive on their own, and the refuge's founding goal of a "self-sustaining population" will have been achieved. There is little evidence that this is actually happening, however.

Chapter 8 turns to the issue of public use on the refuge, in particular bird-watching. The focus of most public use is not the masked bobwhite, but the variety of other birds and wildlife that occur "naturally" on the Buenos Aires by virtue of its location, topography, and vegetation: in short, its habitat. In this connection, the emphasis shifts from the masked bobwhite to what it symbolizes: namely, the refuge as a natural place, with cattle ranching explicitly cast as the unnatural opposite. Both refuge managers and public visitors rely heavily on a timeless notion of "nature" to validate their activities. Imagining the refuge as either restored to, or a remnant of, its "pristine" conditions sanctifies the Buenos Aires as a sacred space, a destination for tourists seeking respite from their everyday urban lives. It also ratifies the notion that removing cattle from public rangelands necessarily benefits wildlife. These claims have little basis in fact, however. They are myths generated by the environmental politics of suburbanization, and they obscure important and difficult scientific questions about the Buenos Aires and the Southwest as a whole.

Chapter 9 addresses the conflict over the Buenos Aires directly. The major challenges for ecological restoration—mesquite encroachment and arroyo formation—have not been addressed on the ground or in public debates. Instead, debate has focused on cattle and the masked bobwhite, each symbolizing a system of value that devalues the other. The conflict, then, is not really about the environment, although it is waged in environmental terms. Rather, the opposition between ranching and the refuge is a symbolic one, born of the structural tension between competing forms of land use and capital accumulation. The Buenos Aires Refuge appears to resolve this contradiction by rejecting both ranching and subdivision in favor of transcendent values of wildlife and Nature. Its size alone lends a sense of significance to this accomplishment. The larger effect, however, has been quite the reverse. On regional and national scales, the refuge has aided and abetted suburbanization by obscuring (or "mitigating") its impacts elsewhere, by appeasing the recreational appetites of its customers and the political demands of its potential critics, and by casting aspersions on those whose lands it seeks to devour, sooner or later. The Buenos Aires has helped drive a wedge between ranchers and environmentalists—perhaps the only coalition capable of containing sprawl and achieving long-term, large-scale conservation in the region. This need not have been the case, and it need no longer be the case.

1

The social production of the endangered masked bobwhite

The Buenos Aires National Wildlife Refuge was established "for the conservation of habitat necessary to begin the recovery process for the endangered masked bobwhite *(Colinus virginianus ridgwayi)*" (FWS 1985; fig. 1). Several additional objectives were developed for the refuge at its creation in 1985, including public use, the protection of other rare flora and fauna, and the restoration of the Sonoran savanna grassland habitat type; in chapters 7 and 8, we will see that these other objectives have become increasingly important over time. The principal justification, however, for the acquisition of the Buenos Aires by the FWS was the masked bobwhite, whose status as a listed endangered species gave the agency authority under federal law to make the purchase. Before turning to the history of the Buenos Aires as ranch and as refuge, it is therefore necessary to examine how the masked bobwhite became an endangered species and how its preservation came to justify, in the eyes of FWS officials and the U.S. Congress, the acquisition of the Buenos Aires Ranch.

My thesis is quite simple: The masked bobwhite, though itself a "natural" product of evolution and a "wild" bird species, became an *endangered species* as a result of *social* processes. Most dramatically, the cattle boom extirpated the bird from Arizona by 1900 (see chapter 2). Likewise, the masked bobwhite's "discovery" and preservation are social phenomena; its elevation to the status of "endangered," no less than its near extinction, was a social process. Over time, the masked bobwhite was invested with symbolic value by a small number of dedicated ornithologists who saw the bird as a victim of excessive livestock grazing. This process was intimately bound up with defining *Arizona's* bird life and with the growing responsibility of state agencies to protect and manage wildlife. Inasmuch as the masked bobwhite was extirpated from Arizona, it was in a sense extinct. It survived only in Mexico, beyond the

Figure 1. Male masked bobwhite in Sonora. In isolating the bird from its environment, the photograph participates in the epistemology of early scientific classification: The organism is defined by its body. To see a masked bobwhite this well in the wild is extremely rare. (Photograph by Steve Gallizioli; courtesy of the Arizona Game and Fish Department)

protection of U.S. laws. Nevertheless, ornithologists and wildlife officials could hold out hope that it might someday be "restored" to Arizona by transplantation. By the 1960s the masked bobwhite's identity was linked to the ravages of the cattle boom and the power of the cattle industry. The importance of drought, and indeed of history, was occluded by the bird's symbolic charge.

If this thesis is surprising, it is only because scientists have tended to emphasize the natural over the social aspects of the bird. Knowledge of the masked bobwhite has generally been confined to a dozen or so amateur and professional naturalists, biologists, and ornithologists, who have focused on the description and analysis of its "natural" characteristics, behavior, and habitat. Although these experts would readily agree that the bird's fate was determined by human activities, they have not made these activities the focus of their research. And with good reason: A narrowly "scientific" approach was critical to securing the legitimacy of their practices, including collecting, identifying, describing, breeding, and releasing masked bobwhites. Somewhat paradoxically, their avoidance of social issues helped to produce the masked bobwhite's symbolic importance and meaning. It was precisely as a symbol of nature and of Anglo-American destruction of nature that the bird could mobilize the political support necessary for the creation of the refuge. But how were these two threads, the scientific and the symbolic, interrelated?

HERBERT BROWN AND THE "DISCOVERY" AND DEMISE OF THE MASKED BOBWHITE

In the late nineteenth century, as the cattle boom overwhelmed southwestern grasslands, Herbert Brown emerged as the masked bobwhite's first student and advocate (fig. 2). Brown grew up in West Virginia and was educated in Cincinnati; he came to Tucson in 1873. He was involved in a lumbering enterprise in the Santa Rita Mountains south of Tucson and then in a mercantile firm (*Arizona Weekly Star*, 5 June 1879, cited in Bahre 1991: 170; Schellie 1970); most sources describe him as involved in mining. Clearly he was a member of Tucson's small but emergent Anglo-American bourgeoisie. In 1885 he became part owner of the *Arizona Citizen*, which he later owned outright; he sold the paper in 1901. For most of the period he was also manager or editor, a position from which

Figure 2. Herbert Brown, "discoverer" of and first advocate for the masked bob-white. (Courtesy of the Arizona Historical Society, AHS no. 11425)

he advocated for "progress" in Tucson and Arizona: statehood, education and business development, railroad expansion, public works, and extermination of the Apaches. In 1898 he took a government post as superintendent of the territorial prison in Yuma (Schellie 1970). Just before his death in 1913, he served as clerk of the Pima County Superior Court.

In his free time, Brown was an amateur naturalist and collector of bird, insect, and reptile specimens, many of which he shipped to eastern institutions such as the Smithsonian for taxonomic identification. Most likely his mining interests prompted him to travel many times through the Altar Valley and into Mexico. Sometime before 1881, in Sonora, he had his first encounter with the masked bobwhite:

> It is not easy to describe the feelings of myself and American companions when we first heard the call *bob white*. It was startling and unexpected, and that night nearly every man in camp had some reminiscence to tell of Bob-white and his boyhood days. Just that simple call made many a hardy man heart-sick and homesick. It was to us Americans the one homelike thing in all Sonora, and we felt thousands of miles nearer to our dear old homes in the then far distant States. The omnipresent hope of "striking it rich" has made life's burden light to many a weary man, and when the 'Perdice' [as the Sonorans called it] made its sweet call only those who have been similarly circumstanced can appreciate it as we did. Then, though but a young man, I had spread my blankets over much of the frontier West, and no one felt that letter from home more than I did. (Brown 1904: 210–211)

Even allowing for the likelihood that twenty-five years of advocacy in behalf of the masked bobwhite had heightened the power of this recollection, the conjuncture of factors that made Brown's "discovery" possible is extraordinary. The bird's distinctive call was embedded in the Americans' memories because the northern bobwhite *(Colinus virginianus)* had long been "America's most familiar game bird" (Brown 1989: 124). With a range stretching from the Gulf of Mexico to the Great Lakes and from the Atlantic Coast to the Great Plains, the bobwhite was indeed a national bird. Most American males of the time, from a young age, hunted it for sport or for food. The bobwhite was unknown from the Rocky Mountains west, however; its call in Sonora at once accentuated and assuaged a sense of being out of place. The dream of riches, Brown suggests, had brought him and his companions far from home; the call of the bobwhite transported them back, at least momentarily. Sonorans, who *were* at home, could not have had this striking experience; Americans could not have had it with a species native *only* to the Sonoran Desert. The masked bobwhites had, sometime in the distant past, split off from their brethren to the east and adapted, becoming at home in the Southwest: a condition to which Brown and his companions could only aspire. For Sonorans (as far as we know) the masked bobwhite was simply another game bird, no more or less remarkable than the three other species of quail found in the region.[1] For the immigrant Americans, however, it was a visceral mnemonic of home, a symbol of uprootedness and loss. Its power, moreover, partly derived from a negative construction of the motivations underlying westward expansion. Brown constructed the masked bobwhite nostalgically, as a symbol of lost innocence and freedom from the press of money making.[2]

Over the ensuing decades, Brown ensured that the existence of this singular bird would not be lost to posterity, though it required a rather protracted struggle waged across the breadth of the nation. He appears to have given the bird its common name, which refers to the black markings on the male's head. Scientists in the East were as surprised as he was, and without having shared his experience, they were skeptical of his claim that bobwhites inhabited Sonora and Arizona. To persuade them, Brown endeavored to secure specimens, obtaining two in 1884 from a man named Andrews, who had killed them on the eastern slope of the Baboquivari Mountains in the Altar Valley. Brown published a note in the *Arizona Citizen* and in the natural-history periodical *Forest and Stream,* describing the birds as *Ortyx virginianus,* the scientific name for northern bobwhites at that time (Brown 1885). But the specimens spoiled before they could be shipped to a more authoritative scientist in the East. R. R. Ridgway of the National Museum (for whom the species was eventually named) published his opinion that Brown had captured Mearns' quail (*Cyrtonyx montezumae*), a known species, the females of which resemble

female northern bobwhites. Later, upon examining the deteriorated remains, Ridgway declared them to be *Ortyx graysoni*, a known bobwhite species from farther south in Mexico. Finally, an intact specimen was examined by William Brewster of Cambridge, Massachusetts, who declared it to be a new discovery and named it *Colinus ridgwayi* (Brewster 1885).

The specimen Brewster examined was the one killed by Frank Stephens in Mexico in August 1884. Very likely capitalizing on the notoriety of the masked bobwhite, Stephens published an account of the trip in an ornithological journal (Stephens 1885). He described having first encountered the birds along the road in the bottomlands of the Altar Valley, sometimes hidden in the thick grass and sometimes perched in nearby mesquite trees. They were so difficult to see or approach that for three days he could not bag one, though he heard them calling frequently and fired on many. (He shot at almost any bird he saw, killing about fifty in three weeks' time.) The type specimen he collected was only the first of 258 that would end up in museums in London, Paris, and throughout the United States, as collectors continued to kill masked bobwhites in Mexico until 1931 (Tomlinson 1972b: 4–5).

Stephens's type specimen still did not fully resolve the scientific dispute, however, because it turned out to be somewhat atypical. It had no white on its head, whereas other masked bobwhite specimens did. Finally, in 1886 J. A. Allen of the American Museum of Natural History in New York published comprehensive descriptions (1886a, 1886b, 1887) based on twenty-one masked bobwhite specimens and several dozen specimens of other known bobwhite species. He concluded that "[t]he species of the genus *Colinus* present a most interesting and puzzling group, consisting of a number of obviously more or less unstable forms, evidently derived, at no very remote period, from some common ancestor" (Allen 1886b: 276). He struggled to establish criteria that would distinguish the masked bobwhite specimens from the others. There were no consistent differences in form or size, and the females of several species were indistinguishable. Only plumage appeared to differentiate the specimens, but it varied widely, especially within the genus as a whole but also within each species. Nevertheless, describing the plumage of the masked bobwhite in excruciating detail, he prevailed in finding it to be a distinct species, while conceding that further research might prove it to be the same as *Colinus graysoni* (1886b: 282; the genus name was changed from *Ortyx* to *Colinus* in 1884).

Logically there was a tautology at work in the classification of bobwhites. Distinctions putatively based in the bodies of the birds themselves, abstracted from space and time, were found to correlate auspiciously with geographical place of origin. This tension would later be resolved by Ernst Mayr's biological species concept, which relocated the foundation of species distinctions in the geography of reproduction rather than aspatial morphology. Accordingly, in 1944 the American Ornithological Union changed the scientific name of the masked bobwhite to *Colinus virginianus ridgwayi*, establishing it as a subspecies.[3] Regardless of this modification, the bird's scientific identity, which is universal, derives from its geographic identity, which is particular. The result is misplaced concreteness: Phenomena that result from spatiotemporally specific relations between an organism and its habitat are misrecognized as essential qualities of the organism itself, abstracted from space and time. The symbolic potential of the masked bobwhite here takes root.

In addition to physical description, scientific classification involved defining the masked bobwhite's geographical range. Herbert Brown's writings contain two judgments on this point. On the basis of collected specimens, he judged the range in the United States as the Altar and Santa Cruz Valleys, not more than thirty-five miles north of the international border. Based on reported sightings and the testimony of old residents, he sometimes extended this U.S. range eastward about one hundred miles, taking in parts of the Sonoita Creek, Babocomari Creek, and San Pedro River valleys and reaching as far north as fifty miles from Mexico. Subsequent researchers have discounted the extended boundaries and added the Arivaca area connecting the Santa Cruz and Altar Valleys (map 2).[4] Either way, the great majority of the masked bobwhite's overall range lay in Mexico; Brown presciently suggested that it extended southward to around Guaymas. "It is doubtless properly a Mexican species," wrote Allen (1886b: 287), applying the curious but customary geopolitical overlay of taxonomy. Nevertheless, Brown's efforts established two facts critical to the masked bobwhite's later protection under the Endangered Species Act: first, that it warranted scientific status as

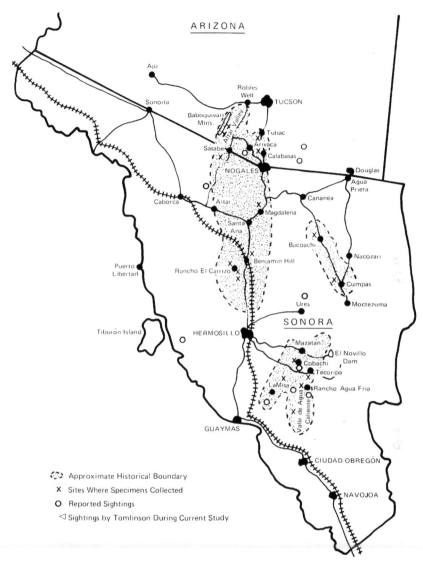

Map 2. A U.S. Bureau of Sport Fisheries and Wildlife map of the historical distribution of the masked bobwhite in Sonora and Arizona, 1884–1900. (Map by Roy Tomlinson)

a distinct (sub)species, and second, that it had inhabited the United States, making it part of "our fauna" (Brewster 1885: 199).

Brown's final achievement in advocating for the masked bobwhite was to link its demise to the impacts of cattle grazing. As cattle numbers grew through the 1880s, he had increasing difficulty locating masked bobwhites. He recorded no sightings in the Altar Valley after 1893; he collected four birds from the Santa Cruz Valley in late 1896 but none in Arizona after that (Brown 1904). In an article published in 1900, he criticized the cattle industry for its impact on bird life. Human settlement was so sparse as to make subsistence hunting a negligible factor, he observed. "Cattle interests are, however, a dominant feature in the development of the country, and to these, more than all else combined, must be charged the obliteration of bird life in the so-called desert portions of the Territory" (Brown 1900: 31). He then offered perhaps the most passionate description of the 1891–93 drought made by anyone living at the time:

> From 50 to 90 per cent of every herd lay dead on the ranges. The hot sun, dry winds and famishing brutes were fatal as fire to nearly all forms of vegetable life. Even the cactus, although girdled by its millions of spines, was broken down and eaten by cattle in their mad frenzy for food. This destruction of desert herbage drove out or killed off many forms of animal life heretofore common to the great plains and mesa lands of the Territory. Cattle climbed to the highest mountains and denuded them of every living thing within reach. Often many miles from water and too weak to reach it they perished miserably. (Brown 1900: 31)

For Brown, who as editor of the *Arizona Citizen* had publicly preached economic development for more than a decade, these were extremely strong words. He did not blame livestock alone; commercial game hunters were also impacting quail populations, capturing thousands of birds for shipment to market in San Francisco.[5] The masked bobwhite, however, was not numerous enough for commercial trapping. By the time legal restrictions on quail hunting were passed, the masked bobwhite was already gone from Arizona.

In what would become a theme in masked bobwhite research for decades to come, Brown concluded: "The causes leading to the extermination of the Arizona Masked Bob-white *(Colinus ridgwayi)* are due to the overstocking of the country with cattle, supplemented by several rainless years" (Brown 1904: 213). He opined that the same combination would soon extirpate the masked bobwhite in Sonora as well. Its habitat—stands of native grasses in broad valleys—placed it in direct competition with cattle. If the bird had survived in Sonora, he reasoned, it was only a matter of time before cattle numbers and drought would take their toll. A key factor in this process was the advent of wells for watering stock. Areas near natural waters were already heavily grazed; artificial waters enabled cattle to graze in more remote areas where grasses were relatively healthy: precisely those areas most likely to harbor remnant populations of quail. Brown's last published words on the masked bobwhite—"Unless a few can still be found on the upper Santa Cruz [in Mexico] we can, in truth, bid them a final good-bye" (1904: 213)—suggest that he believed the bird was extinct when he died in 1913.

J. STOKLEY LIGON AND EARLY RESTORATION EFFORTS

In Sonora, however, cattle numbers and water development lagged behind Arizona. In 1937 J. Stokley Ligon and David Gorsuch located and captured 132 masked bobwhites about seventy-five miles southeast of Hermosillo. Ligon (fig. 3) developed what he later described as an "intense interest" in the masked bobwhite, partly because of the lack of solid evidence as to its existence and range (Ligon 1952). From 1937 to his death in 1961, he would carry the mantle first borne by Herbert Brown.

Ligon grew up in Texas, where his father owned a sheep ranch. In 1900, at age 21, he rafted down the Pecos River to the Rio Grande and thence to the Gulf of Mexico, hunting game and observing wildlife. After studying biology and natural history at Trinity University in Texas, he went to work for the U.S. Biological Survey, where his first assignment was a study of waterfowl breeding in New Mexico. Ligon excelled in government service, eventually attaining the post of head field agent for predatory animal control in several western states and Alaska. According to the dust-jacket biography for his posthumous 1961 book, *New Mexico Birds and Where to Find Them,* Ligon "warred on wolves, killed the last

Figure 3. J. Stokley Ligon, first man to breed masked bobwhites in captivity. He made numerous attempts to reintroduce birds—both wild caught and captive bred—at locations in New Mexico and Arizona, but none was successful. The birds in the photograph are Mearns' quail. (Courtesy of the New Mexico Game and Fish Department)

lobo of record, [and] bagged the biggest grizzly in New Mexico." In 1930 he retired from government service to establish an experimental game bird farm near Carlsbad, New Mexico, which was later acquired by the state's Game and Fish Department. His most active masked bobwhite work occurred during this period. He also worked as a collector for scientists and museums.[6] Later he returned to government as a field biologist and a regional supervisor for the FWS.

Ligon's career encapsulates the attitudes and approaches to wildlife prevalent in the early decades of the twentieth century. He believed that human activities had disrupted the balance of nature and that public wildlife management was necessary to restore it (Ligon 1946: 3). With typical Progressive Era confidence, he viewed government conservation as both reversing the negative effects of economic exploitation and ensuring the long-term conditions for human prosperity:

> The encroachment of economic conquest on the wilderness was inevitable in the progress of civilization; and as man has destroyed the natural balance, he must assume the responsibility of establishing and maintaining an artificial one. He must now combat disease . . . and control predatory animals if he would have the increase in game; he must control forest fires and avoid range abuse and erosion if he would hold the soil and grow forage; and, finally, he must utilize surplus game if he would protect it from over-production. (Ligon 1927: 11)

Ligon divided the animal world into categories that were simultaneously economic, political, and moral. Like most conservationists of the time, he viewed game species, livestock, and songbirds as "good" because they yielded economic and moral value; predators, varmints, and rodents were "bad" (cf. Dunlap 1988). He transposed the bureaucratic classification of federal and state agencies onto the world of wildlife, publishing *Wild Life of New Mexico: Its Conservation and Management* in 1927 as well as *New Mexico Birds*.

Publication in 1931 of Herbert Stoddard's classic monograph, *The Bobwhite Quail: Its Habits, Preservation, and Increase,* strongly shaped Ligon's views of the masked bobwhite. Based on five years of field research in

Georgia and Florida, Stoddard's book demonstrated the potential of managing wildlife, especially upland game birds, as a crop that could be maximized through habitat manipulation guided by the insights of science. A scientist in the Bureau of Biological Survey, Stoddard was recruited and supported by a group of well-to-do southeastern sportsmen and hunting plantation owners who desired to increase the quality of the hunt; despite this elite origin, the study's results were construed as being of preeminent public value. The book itself may be likened to a functionalist ethnography of *Colinus virginianus*. How the bobwhite nests, breeds, feeds, sleeps, raises its young, and dies are all systematically presented, in the present tense, as functions of ecological adaptation and survival of the fittest.

Although Stoddard identified economic exploitation—both livestock grazing and modern farming—as the chief threat to game birds, he proclaimed that the bobwhite would be saved by a partnership of market forces and state regulations. "[T]he bobwhite is too valuable an asset to be ignored," he observed (1931: 5). If hunters were charged fees and regulations were enforced against poachers and "gunners" (hunters who were not "sportsmen"), the disrupted "natural" balance would be restored. Not without casualties, however: The Cooper's hawk *(Accipiter cooperi)*, whose predatory activities had helped to cull unfit quail in the past, would have to go. "As hunting by man serves much the same purpose . . . the services of the hawk can be dispensed with to advantage under present-day conditions. Like the picturesque pirates of old, Cooper's hawk is too violent and bloodthirsty to be willingly tolerated" (1931: 212).

If Herbert Brown's relation to the masked bobwhite was mediated by collection and taxonomic description, Stokley Ligon's turned on tools and techniques pioneered by Stoddard: the live trap, captive breeding, and transplantation. Ligon was directly or indirectly responsible for almost every early attempt to reintroduce the masked bobwhite to the United States. His 1946 manual for wildlife managers, *Upland Game Bird Restoration through Trapping and Transplanting*, describes methods of capturing, handling, and releasing quail, grouse, turkeys, pheasants, and partridges, including methods he presumably used with the masked bobwhite.

For Ligon, capturing and transplanting game birds was a meticulous, almost clinical undertaking. It should be done, he wrote, only by trained and experienced professionals "with intimate knowledge of the various species, their habitat requirements, the seasons for operating, care of birds, methods of release, and a multitude of other important details" (1946: ix–x). Birds that have experienced hunting pressure behave differently than do those unfamiliar with the gun, he noted, and captive-reared birds are altogether different in their habits from wild-caught ones. Bodily discipline was key: Humans are game birds' greatest fear, and handlers must carefully control their noise, movements, and visibility to minimize mortality. Ligon provided precise instructions for constructing drive nets, transport crates, and other equipment specific to each species of game bird.[7] He stressed that selection of proper habitat—preferably within the species' historical range—was of paramount importance because food and cover were the birds' greatest needs. "Incompetence in any of these matters, and particularly the last, can result only in failure and in unfavorable public reaction" (1946: 6).

Ligon's manual throws into relief the curious status of the masked bobwhite at the time: In some ways wild and others domesticated, it was technically a game bird but nowhere hunted as game. In the eastern United States at the time, some states were responding to declining populations of northern bobwhites by reclassifying them as songbirds and thereby bringing them under hunting bans (Dunlap 1988: 65). The larger trend, however, was "toward managing upland game birds as a renewable crop to be cultivated and utilized" (Ligon 1946: 4). Perhaps he was optimistic that masked bobwhites would someday be numerous enough to hunt, but Ligon never succeeded in transplanting any to the United States. The birds he captured in 1937, along with some of their captive-bred offspring, were released over the next four years at ten locations "in what was believed to be the best habitat in Arizona and New Mexico" (Ligon 1952: 49); none survived (Brown and Ellis 1977: 8; Tomlinson 1972b: 14–15). Another trip to Sonora in 1949 yielded no birds and heightened fears that the habitat there was deteriorating.

After extensive searching in 1950, Ligon and two partners finally found thirty-four birds at two locations east and south of Hermosillo; they brought twenty-five of them back to the United States (Ligon 1952).

Fifteen were released on Fort Huachuca, a military base about seventy-five miles east of the Altar Valley. As with most of the early releases, this was inside the extended range specified by Herbert Brown but outside what is today considered the masked bobwhite's historical range; the base was selected because no cattle were there. Again, the birds failed to reproduce. The remaining ten became breeding stock at Ligon's Carlsbad game farm (Gallizioli et al. 1967: 572). The descendants of these birds would briefly form the nucleus of the FWS breeding program in 1966–68. By that time, however, it would be hard to consider the birds fully "wild" by Ligon's own arguments.

As in the case of Herbert Brown, Ligon's interest in the masked bobwhite evolved into an animosity toward cattle ranching, which the failure of his transplant efforts only aggravated. The ubiquity of cattle throughout the bird's historical U.S. habitat was a conspicuous obstacle to his ambitions, and the only one clearly subject to human control. Echoing Brown, he concluded that "[e]xcessive range use by livestock and recurrent drought, direct causes of the bird's final extirpation from Arizona, seemingly prevented success" of his early release experiments (Ligon 1952: 49). In his 1946 manual, he observed that "excessive grazing by livestock has resulted in serious depletion and contraction of ancestral ranges" for the masked bobwhite, and he ruefully pointed out that stock tanks on ranches did not help the birds because "overgrazing and trampling by livestock generally destroy ground cover in the vicinity of water" (1946: 9–10). According to a friend who knew him in the late 1950s, Ligon also came to regret his own earlier work in predator control, which had been performed largely in behalf of the livestock industry. "He used to say there aren't a thousand cows worth one wolf."[8]

By the end of his life, in writing New Mexico Birds, Ligon had become more strident. He chose to include the masked bobwhite in his book, writing rather disingenuously: "There seems to be no conclusive evidence that the Masked Bobwhite is a native species of New Mexico. However, its ancestral range, as known to ornithologists, is fairly close to New Mexico, in southern Arizona. It is quite probable that it formerly occurred in the Animas Valley of Hidalgo County, where there is still limited habitat comparable to that of its original Arizona range." On this basis, he saw fit to offer a piece of embittered biography:

In 1937, 1949, and 1950, the author, in company with Arizona Game Department officials, made trips to Sonora, Mexico, when a number of the birds were secured for propagation and stocking in both Arizona and southwestern New Mexico. A "planting" was made in the Animas Valley, south of Animas, New Mexico, and for a time the birds thrived. Because of devastated rangelands, however, there is some doubt as to whether any remain. There is little doubt that this interesting Quail, under more favorable land-use practice, permitting of more normal growth of vegetation, can be established in some parts of the southern and southwest sections of the state. Because of denuded habitat throughout its ancestral range, it is now near the brink of extermination. The main hope in saving it from extinction now seems to rest on restored habitats and propagated birds for stocking. It should be kept apart from the common Bobwhite to avoid hybridization. (Ligon 1961: 95)

The masked bobwhite belonged among "New Mexico's birds" solely because Ligon had put some there. Given the elevation and climate of the Animas Valley south of Animas, it is exceedingly unlikely that the masked bobwhites released there could have survived a winter regardless of grazing pressure. Judged against the principles expounded in his 1946 manual, moreover, releasing the birds there—outside of their ancestral range and in what he later called "denuded habitat"—was guaranteed to fail. It did not provoke "unfavorable public reaction," however. The released birds became martyrs: In dying, they helped the masked bobwhite become a symbol of the impact of cattle grazing on wildlife in general.

THE LEVY BROTHERS AND THE
REDISCOVERY OF WILD MASKED BOBWHITES

Following Ligon's disappointing 1949 and 1950 trips to Sonora, the masked bobwhite was again feared to be extinct, or nearly so, in the wild.[9] Ligon's game farm continued to supply captive-bred birds for a handful of release experiments, but not one was successful (Brown and Ellis 1977). The small group of ornithologists and wildlife experts concerned with the masked bobwhite attributed these failures to the poor quality of habitat in the United States, all of which was grazed by cattle.

As Sonoran rangelands were perceived to be at least as degraded as those north of the border, it followed that prospects for any remnant populations in the wild were dim. Herbert Brown's prediction appeared to have come true.

Fear that it was extinct heightened the masked bobwhite's symbolic value, with the bird increasingly portrayed as a victim of human recklessness. In this construction, the masked bobwhite's relation to the state of Arizona always played a significant part. As one biologist involved in later recovery efforts put it, the masked bobwhite became "Arizona's passenger pigeon,"[10] a comparison that implies extinction. The authors of *The Birds of Arizona,* published in 1964, dubbed it "Arizona's most famous bird," and they chose for the book's frontispiece a large color rendering of a nesting pair, with Baboquivari Peak in the background. In the text, they blamed the bird's demise solely on cattle, making no reference to drought:

> Extinct in Arizona for many years; before 1890 common in tall grass prairies from Baboquivari Mountains east to upper Santa Cruz Valley. Grazed out of existence by early 1900's. . . . Attempts at reintroduction have been unsuccessful, as there is no ungrazed grassland within the former range in Arizona. . . . From time immemorial it had thrived in the prosperous grasslands of the border; but it died off almost instantly at the demise of its home with the coming of the great herds and their owners. Let those who really wish to conserve our wild heritage ponder well the lesson! (Phillips et al. 1964: 28)[11]

Efforts to release masked bobwhites outside of their historical range were a waste of money, the authors insisted; the only hope was to restock the quail in "suitable grasslands, if such can be found or grown" (ibid.: 29). The opposition between cattle and quail had cemented into a rigid dualism.

As it turned out, however, the masked bobwhite was not extinct in the wild. In 1964 Jim and Seymour Levy made a trip to Sonora to look for the bird. The Levy brothers had grown up in Chicago, where they learned bird-watching from a mentor who worked as a grave digger in a suburban cemetery. After serving in World War II, they returned to Chicago, later moving to Tucson for their parents' health. Taxidermy, earlier a hobby, became their living; advocating for wildlife became their avocation. They met Stokley Ligon in the late 1950s. Shortly before Ligon's death, they acquired four or five pairs of his masked bobwhites and learned how to propagate them in captivity. Fearing that the captive birds were becoming inbred, they resolved to search Sonora for wild populations. On a 1963 trip, Jim Levy surveyed all the locations where Ligon and others had found birds; he found one feather near Guaymas, but no live birds anywhere. Having given up the search for bobwhites, he stopped thirty miles north of Hermosillo for lunch and was surprised to find elegant quail there.

The following year, on a return trip with Seymour and AGFD biologist Steve Gallizioli, Jim Levy pulled over to show his brother where he had found the elegant quail, but he mistook the location. This time he was farther north, on the Rancho El Carrizo near Benjamin Hill. As before, they had found no masked bobwhites farther south. Looking for elegant quail feathers, Levy dismembered a cactus wren nest and found instead a masked bobwhite feather. A search of the area turned up a population of live birds. By all accounts the discovery was accidental,[12] though Levy recalls that the area around the nest was conspicuously better habitat than most of the surrounding country. It stood out, he recalls, "like a fire hydrant in a parking lot." The area supports what is now taken to be the last remaining wild populations of masked bobwhites.

The Levys' discovery renewed hopes that the captive breeding program could succeed long enough to reintroduce masked bobwhites to Arizona. Government agencies began to play a greater role than previously, but efforts were constrained by a lack of funding. In the United States, under statutory principles derived from English common law, wildlife is held to be public property. By analogy with commerce, wildlife that crosses state boundaries (such as migratory birds) is the legal jurisdiction of the federal government, whereas intrastate wildlife falls under state jurisdiction.[13] Federal agencies had helped Ligon obtain permits to collect birds in Mexico (Ligon 1952: 48), but as a nonmigratory species, the masked bobwhite was a state responsibility. The New Mexico and Arizona Game and Fish Departments had shown interest before, but mostly through their game programs; little or no funding was earmarked for the masked bobwhite because it was not a quarry for hunters. Participation was generally limited to staff time and intergovernmental consultation.[14]

When the Levys attempted to secure protection for the remaining birds in Sonora, there were expressions of support from government agencies on both sides of the border, but no state funds were forthcoming, and the effort foundered when the landowner backed out (Gallizioli et al. 1967: 574). The Levys then initiated a correspondence with South Dakota Senator Karl Mundt, who was sponsoring the Endangered Species Act. Mundt got funding in 1965 for the FWS's Endangered Wildlife Research Program, with the masked bobwhite as one of its first projects (Ellis and Serafin 1977: 21; Ricciuti 1979). The program was soon subsumed under the Endangered Species Preservation Act.

SPECIES OF CAPITAL
THE LISTING OF THE MASKED BOBWHITE

The Endangered Species Preservation Act of 1966 changed the entire context of masked bobwhite preservation. Although concern for certain wildlife species—bison, deer, eagles, and waterfowl—was longstanding, and game laws had existed since colonial times, the act mandated that *any* threatened or endangered species be protected, and it authorized federal expenditures for that purpose. As amended in 1969, the law called for the creation of an official list, overseen by the FWS, of flora and fauna deemed at risk of extinction. As one of the first "listed" species, the masked bobwhite suddenly became a priority for federal agencies, in particular for the FWS. According to Jim Levy, "Nobody was interested in that bird in the federal government until the Endangered Species Act, when they had some money for it" (personal communication, 20 March 1998).

Under the Endangered Species Act, the Linnaean classification system of biology became a legal and bureaucratic one: the official framework for identifying and managing certain plants and animals. The abstraction inherent in the Linnaean system was no longer merely a scientific construct but a presupposition of agency work, both in the lab and in the field. The masked bobwhite's life history, historical range, and genetic specificity became the basis of research and preservation efforts. By making endangered species preservation a funded mandate, the Endangered Species Act altered the relationship between endangered wildlife and other fields of government activity, and therefore the relationship between the FWS and other agencies (Bourdieu 1998: 42). Though less famous than the whooping crane and the California condor, the masked bobwhite suddenly had a direct legal claim to federal funds (Ricciuti 1979); its fortunes would henceforth serve as one measure of the success of the FWS and its programs. The steps taken after 1966 have a systematic logic reflective of these new circumstances.

First, the FWS became the exclusive guardian of the birds, asserting a sort of monopoly property right. It became illegal for a private citizen to possess a masked bobwhite without obtaining FWS permission. The Levys donated the last of Ligon's masked bobwhites (four male-female pairs) to the newly created Patuxent Wildlife Research Center in Maryland in 1966 (Brown 1989: 135; Ellis and Serafin 1977: 23; Tomlinson 1972a: 296). Low breeding rates with these birds prompted the FWS to secure wild birds from Rancho El Carrizo. Biologist Roy Tomlinson captured thirty-six masked bobwhites in 1968 and another twenty-one in 1970 for relocation to Patuxent (Tomlinson 1972a: 305). These resulted in markedly higher reproduction rates, and the offspring of Ligon's birds were culled (Ellis and Serafin 1977: 23; Tomlinson 1972b: 16). All captive birds and their offspring were banded for individual identification.

Second, captive breeding techniques were perfected to maximize offspring and prevent inbreeding. Adapted from the commercial poultry industry, these practices represented a new level of rationalization and intensiveness:

> Newly hatched masked bobwhites are raised in commercial-type, electrically heated, thermostatically controlled battery brooders with raised wire floors in a building with continuous forced draft ventilation held at a temperature of approximately 75 degrees F. The brooding temperature is maintained at 105 degrees F during the first two weeks of age. Chicks are lightly debeaked following hatching to prevent cannibalism. When properly debeaked at an early age, the beak regrows normally before birds are released. The young are supplied a finely ground quail starter ration and water on an *ad libitum* basis. The principal ingredients include corn, soybean, fish, and alfalfa meals. Beginning about 3–6 weeks of age the diet is supplied in the form of $3/32$" pellets. (Ellis and Serafin 1977: 23)

The starter ration consisted of seventeen ingredients, formulated with precision down to one ten-thousandth of one percent. The addition of antibiotics to the starter ration was found to reduce chick mortality dramatically. Adult birds' rations included balanced amounts of forty nutrients. These practices persist in masked bobwhite propagation to this day. Light and temperature are monitored for all birds; "[b]reeding birds are provided with [i.e., forced to endure] continuous light from mid-April to late August" to stimulate breeding (Ellis and Serafin 1977: 23–26). When humidity was found to determine breeding behavior, it too was controlled. A computer program was developed to minimize inbreeding.

By necessity, Patuxent staff produced a "total institution," a completely artificial environment to meet the "natural" needs and dispositions of the masked bobwhite, and in this they were quite successful. In 1970 and 1971 they shipped more than six hundred chicks to Arizona for release. In 1976 the program produced more than sixty-eight hundred eggs, and some thirty-nine hundred chicks were hatched (Ellis and Serafin 1977: 23; Tomlinson 1972a: 307).

Third, a comprehensive survey of wild populations was undertaken in Mexico. From 1967 to 1970, Roy Tomlinson searched previously known and promising habitat in Sonora. He found birds at only two sites: Benjamin Hill and an area east of Mazatán (Tomlinson 1972a); the latter population could not be relocated in 1974 and was presumed "extirpated by heavy grazing" (Brown 1989: 128). At Rancho El Carrizo, Tomlinson conducted a study of temperature, rainfall, humidity, and vegetation, as well as the population, life history, and habits of wild masked bobwhites. He learned that the masked bobwhite, unlike other bobwhites, does not begin breeding behavior until July or August, with the onset of summer rains; the males "do not begin to call until minimum daily temperatures and relative humidity average at least 55 degrees F and 25 percent, respectively" (1972a: 303). He also discovered that hatching peaks in mid-September, at the peak of vegetation growth and insect abundance.

Research into these questions continues to this day, but all studies of the wild population have been of uncertain applicability to reintroduction efforts in the United States for a simple but fundamental reason. As Tomlinson (1972a: 307) put it, "[N]o area within the past Arizona bobwhite range was found to be exactly comparable to that in Sonora. Generally, most of the Arizona habitat is higher in elevation (2,300–4,000 feet) than the Sonoran habitat (800–2,500 feet). In addition, the terrain in Arizona is rockier and the vegetation is somewhat different." Whatever the characteristics of pre-1890 Arizona habitat were, they cannot be inferred in any simple way from the habitat in Mexico.

Fourth, in 1969 the FWS launched studies to identify potential reintroduction sites in Arizona. The difficulties of this task were substantial, as Tomlinson (1972a: 307) recognized: "Because a life history study had just begun, not enough was known about masked bobwhite habitat requirements to make the best possible evaluations. The major criterion in selecting areas, therefore, was whether a specific location fell within the historical range of the subspecies. . . . Although selected areas were somewhat less than ideal for the intended purpose, they represented the best available at that time." The FWS attempted to heed Ligon's counsel to evaluate release sites carefully, but without a firmer idea of how to evaluate them, the FWS was forced to decide primarily on the basis of historical range: hence, the Altar Valley.

CONCLUSION

Herbert Brown did not discover the masked bobwhite; the bird had been known to humans for centuries before. He did *document* its existence, however, making it official by enabling scientists to assign it a place in the universal taxonomy of fauna. A striking conjuncture of circumstances made this possible and predisposed Brown to view the bird symbolically, elevating its unique scientific identity to a second order of meaning and value. In this process, the abstract quality of scientific classification reinforced a construction of nature apart from and opposed to human history and society. Stokley Ligon embraced this dichotomy explicitly, and he saw in public wildlife management an enlightened means to rebalance Nature. By linking the masked bobwhite to Arizona, he and his contemporaries made restoring the bird a metonym for healing the state's environmental wounds, particularly those related to cattle. Thus the masked bobwhite came to symbolize innocent Nature despoiled.

The symbolic value of the masked bobwhite was heightened in the

1950s and early 1960s by the widespread notion—half fear, half belief—that it was already extinct in the wild. The repeated failure of reintroduction experiments (most using captive-bred birds) drew attention to the lack of ungrazed habitat in the United States, and scientists began to omit the role of drought in the bird's earlier demise. When the Levy brothers "rediscovered" wild birds in 1964, fear of extinction transformed into ardent hope that "Arizona's passenger pigeon" might yet be saved. Listing of the masked bobwhite as an endangered species ratified its symbolic value and made available significant public monies for the first time.

Captive breeding of masked bobwhites was easier than with most endangered species, thanks in large part to methods and technologies developed for poultry. Reestablishing a "wild" population, however, was (and remains) very difficult. The selection of the Altar Valley for release experiments in the 1970s, based on a map of the masked bobwhite's historical range, assumed that habitat suitability was in some way a timeless feature of the place; the Endangered Species Act contained the same assumption, insofar as the law applied to the masked bobwhite at all. This notion, however, contradicted the historical fact that the bird had been extirpated *because* its habitat had changed. It was like looking back in time through a telescope: Everything else disappeared from view but the singular, magnified image of the masked bobwhite inhabiting the valley in the 1880s. The intervening history was overlooked, while dynamic relations between animal and habitat were reified as timeless, "natural" attributes of the species and the place. I will return to the release experiments in chapter 6. First, it is necessary to reconstruct the missing history of the Altar Valley to understand how the masked bobwhite's erstwhile home had changed.

2

The cattle Boom in the Altar valley

The tenth U.S. census, conducted in 1880, found Don Pedro Aguirre Jr. living in Arivaca. The town was older than the census itself, having been established in the early 1700s by the Jesuit missionary Eusebio Francisco Kino. The site had doubtless attracted Native American use for many centuries before. It marked the point where water from two mountain ranges collected in a small basin before draining west down Arivaca Creek toward the Altar Valley. Reliable springs supported a large *ciénega,* or marsh, rich in grasses and sedges and lined with tall cottonwoods, ash, hackberry, and mesquite trees. In the semiarid landscape, Arivaca was an oasis.

Aguirre was forty-five years old, a successful businessman, and recently elected to the Pima County Board of Supervisors (fig. 4). The honorific "Don" bespoke his Spanish aristocratic descent and his prominence in the tight-knit elite of northern Sonora and southern Arizona Territory. For most of his adult life he had prospered from a stagecoach shipping business started by his father that carried freight and passengers between Tucson and points as far away as Missouri. Many of his routes extended south and west, servicing mining settlements and towns on both sides of the international border. He used the proceeds from the business to invest in mining ventures and real estate and to promote various civic causes. The year before, Arivaca's first schoolhouse had been completed at Aguirre's expense.

Despite his successes, Aguirre faced several difficult challenges in 1880. The Arivaca Mining District, founded just three years earlier, had recently collapsed, with most miners migrating to the richer deposits around Tombstone. Aguirre's long-distance freighting contracts were threatened by the Southern Pacific Railroad, which would complete its transcontinental line through Tucson the following year. That would leave the family business almost wholly dependent on mining, and thus

Figure 4. Don Pedro Aguirre Jr., founder of the Buenos Ayres Ranch. (Courtesy of the Arizona Historical Society, AHS no. 1826)

exposed to the industry's notorious booms and busts. By 1886 Aguirre would put the business up for sale.

Increasingly, Aguirre was turning to livestock for his livelihood, and it was for this reason that the census took notice of him in its supplemental report on cattle, sheep, and swine (though it misspelled his name as "Aguené"). In the late 1870s he had begun to pasture sheep on contract for other owners, including the Tucson firm Lord and Williams and Territorial Governor A.P.K. Safford. Newspapers reported he had "thousands" of sheep in 1876, some ten thousand in 1877, and fifteen thousand sheep plus some cattle in 1879 (*Arizona Citizen,* 19 August 1876, 4 and 18 August 1877; *Arizona Daily Star,* 4 September 1879). Many of these sheep, it appears, had been trailed to Arizona from California, where drought had crippled the range in the early 1870s (Kinney 1996). According to the 1880 census, Aguirre had "some 10,000 sheep, together with about 400 Mexican cattle" and was by far the largest operator in the Arivaca Valley (U.S. Department of the Interior 1883: 95). In the fall of 1881 Aguirre sold three hundred cattle to the San Carlos Indian Agency (*Arizona Citizen,* 2 October 1881).

Even in livestock, however, Aguirre faced a difficulty: securing rangeland for his expanding flocks and herds. More accurately, the problem was water. Large tracts of the public domain remained unclaimed and open to grazing, but virtually every natural water source had been claimed and put to use for livestock. In 1877 Aguirre had filed on 640 acres of the rich lands around Arivaca under the Desert Land Act, but the claim was disallowed pending resolution of the Arivaca Land Grant, a larger claim left over from the Spanish period (Mattison 1946; Wagoner 1952: 64). The census report described the situation: "While the adjacent country of broken mesa lands and foot-hill areas is more or less grassed with the common forage plants, only the country immediately in the vicinity of the streams is used by the ranchers of the [Arivaca] valley, whose herds graze in hot weather chiefly on the moist lands." The valleys to the east had more water sources, but there, too, "such grazing lands as are convenient to water have long since been fully stocked" (U.S. Department of the Interior 1883: 95).

There was abundant grass just west of Arivaca in the Altar Valley. Aguirre knew the area well, having traversed it many times in his shipping business. In 1859 he had established a stage stop on the opposite side of the valley, at the mouth of Presumido Canyon near the international border. The site was on the main road from Tucson to Altar, Sonora, and Aguirre used it to keep fresh horses and mules for his stages. At first he called it La Posta de Aguirre; later he changed the name to Buenos Ayres because of the steady winds there (Leavengood 1993; the spelling was changed to Buenos Aires by a later owner). He abandoned the site in 1864 and moved to Arivaca, apparently in response to the increase in Apache raids during the Civil War.[1] Insecurity persisted for several years after the war. Aguirre's brother was killed en route from Altar to Arivaca in 1870 (Aguirre 1969). By the late 1870s, however, the principal obstacle to grazing in the Altar Valley was the lack of water sources. Very likely, Aguirre grazed the sheep in his care far into the valley because sheep can go farther from, and longer without, water than can cattle. His cattle probably remained in the Arivaca area.

In the decade of the 1880s, Aguirre made the Altar Valley's grasses available to cattle by developing artificial water sources. Others followed his lead. All of them were reacting to the same opportunity, created by the combination of three factors: the natural abundance of forage on vast tracts of public land, government policies that left that land open to all comers, and a surplus of outside credit for aspiring livestock producers. A pattern that had already unfolded across the Great Plains would soon repeat itself in the Altar Valley and in southeastern Arizona as a whole.

THE CATTLE BOOM

In the three decades after the Civil War, profit-driven range cattle production swept north and west out of Texas and across the Great Plains, the intermountain region, and the southwestern deserts. It was a feverish, speculative rush for free grass on the vast public domain. The resulting damage was so profound and enduring that the environmental effects of grazing today cannot readily be separated from those of the boom period, especially in the Southwest. Hence the biologists' judgment, mentioned in the introduction, that cattle grazing is destructive.

The historians' conventional notion of ranching reveals its limitations in attempting to understand and explain the cattle boom, especially in regard to its ecological consequences. Terry Jordan, for example, blames the boom's effects on a cultural "maladaptation" to arid environments (1993: 236–237), even though the tremendous growth in cattle numbers was exogenous to the cultural system he describes. Others have focused on the arrival of the railroad, which enabled the integration of western rangelands into national and international markets for cattle and beef (Webb 1931). The railroad by itself, however, cannot be viewed as the cause of overgrazing; it was a necessary condition but not a sufficient one. Another theory sees the cause in "cultural factors" characteristic of Anglo-American civilization. The Anglo-American "belief system" commodified nature, viewing it "primarily as raw material for wealth" (Hirt 1989: 180). This may be true, but seeing something as a commodity is not tantamount to exploiting it as a commodity. Moreover, cattlemen were themselves among the first to notice and warn against overstocking (Bahre 1991; Griffiths 1901; Hastings and Turner 1965: 41).

The force behind the cattle boom was surplus capital generated by imperialism and the Industrial Revolution in Great Britain and the northeastern United States (Sayre 1999). Staggering profits from cheap cattle and free grass triggered a flood of outside capital onto the western range, especially after the panic of 1873 devalued assets and caused smaller, western banks to fold (Atherton 1961). Operations without significant backing could not survive the droughts, blizzards, and market fluctuations that characterized the industry throughout the period.

Historians have estimated that between 1880 and 1884, "[p]erhaps as much as 15 percent of the total British investment in western America was in ranching"—a total of roughly $45 million for the decade as a whole (Frink et al. 1956: 319, 223). "During the heyday of open-range ranching in the 1870s and 1880s, the industry probably had a higher proportion of European investment than any other western business" (White 1991: 261). American investments were larger still. Working from incorporation records, Gressley (1966: 105) concluded that more than $284 million was invested in cattle operations in Montana, Wyoming, Colorado, and New Mexico between 1880 and 1900. "A perusal of dozens of ledgers and hundreds of letters between bankers and cattlemen leads to the inescapable conclusion that the Western range cattle industry during the last two decades of the nineteenth century was operated basically on borrowed capital" (Gressley 1966: 145). The cattle boom quite simply would not have happened—at least not on the scale that it did—without this massive influx of capital from outside the region.

What attracted the flood of capital? As a British Royal Commission concluded in 1882, "the American stockman of the West is possessed of singular advantages; land for nothing, and abundance of it" (cited in Frink et al. 1956: 141–142; cf. Graham 1960; Gressley 1966). Knowing that they did not own the lands on which their cattle grazed, "every man was seized with the desire to make the most that was possible out of his opportunities while they lasted. . . . There was no rent to pay, and not much in the way of taxes, and while these conditions lasted every stockman thought it well to avail himself of them. Therefore all bought cows to the full extent of their credit on a rising market and at high rates of interest" (Bentley 1898: 8). The resulting debt compelled operators to disregard ecological limits, particularly when drought undermined range and market conditions simultaneously (Abruzzi 1995).

By most scholarly accounts, the cattle boom crested in 1885, when overstocking in the Great Plains led to disastrous die-offs of animals and huge financial losses (Osgood 1929; Webb 1931). Drought in the southern plains killed thousands of cattle in 1883–84, and owners shipped as many animals as they could to pastures elsewhere, including both the northern plains and the Southwest. More animals perished, in turn, in the blizzards that blighted northern states in the winter of 1886–87. By 1888 "much of the western ranching industry was lying in ruins, the victim of severe overgrazing and desperately cold winters. Many thousands of animals were lying dead all over the range, starved and frozen; the survivors were riding in boxcars to the stockyards for rapid liquidation by their owners" (Worster 1992: 41). In Arizona, however, the damage was just beginning. Faced with large herds of animals, low market prices, no forage, and high debt, operators had little choice but to default or move, and Arizona was one of the last places where free grass remained.

DEVELOPING WATERS

In the Altar Valley, in the pediment zone up against the Baboquivari Mountains, several homesteaders had dug shallow wells in sandy washes and established small stock operations before 1886 (Roskruge 1885–86). The wells, however, yielded only small quantities of water, insufficient to support large herds of stock. Down in the central valley, early attempts to dig wells were in vain because the depth to water was too great. "Poor men have spent their savings of years in efforts to obtain water by digging," wrote a journalist for the *San Francisco Bulletin* (reprinted in the *Arizona Daily Star,* 10 February 1887). Holes dug in the loamy bottomland would at best capture water from floods, which might last a month or two. This may have been the origin of some *charcos,* or pools, reported by later observers, who took them to be natural formations (Bryan 1920; Dobyns 1981; Pumpelly 1870: 38).

Aguirre was the first person to develop artificial waters in the central valley, beginning in 1883. He began by constructing a large reservoir near the head of the watershed at the confluence of the Lopez and Compartidero Washes. One imagines him studying the landscape while herding sheep in the 1870s, for his choice of location was expert. The reservoir, which came to be known as Aguirre Lake, lies in lower alluvium just below the pediment zone. There it captured the runoff from the pediment's shallow soils. Because the bottomlands were nearly flat, even a low dam could back up water over a considerable area. The soft soil was easily dug up and moved with *fresnos* (large shovels pulled behind teams of mules or horses). Excavated dirt probably served as raw material for the adobe bricks used to build the headquarters residence, which Aguirre located on a hill just south of the lake.

The summer of 1885 was unusually dry, and the lake did not fill. This may have compelled Aguirre to invest in drilling a well. In the spring of 1886 Aguirre struck water at 515 feet below the surface. Even the well did not guarantee water, however: The *Arizona Daily Star* of 7 May 1886 reported that the pump had broken a valve and would need repairs. Aguirre may also have developed some stock tanks: smaller dams pushed up in tributary drainages to capture runoff. In the township surrounding the headquarters (an area six miles by six miles), more than two dozen such tanks were constructed between 1883 and 1919, with water rights to almost 120 acre-feet of water per year.[2] In the absence of these constructed waters, the rich grasses of the Altar Valley could never have sustained large herds of cattle because of the excessive distance to water.

Aguirre's water improvements served the additional purpose of helping to secure control over the range. It was illegal to fence the public domain, but under customary "range rights," the owner of a water source had presumptive right to graze halfway to the next source. By this measure, Aguirre Lake alone probably bestowed control over some thirty thousand acres of land. Improvements were also necessary to secure title to land under federal law. The Desert Land Act permitted claims of 640 acres in arid regions on the condition that irrigation be developed. The act is infamous for the fraud it spawned. Many livestock operators filed claims simply to secure grazing land for three years at twenty-five cents an acre, with no intention of ever making the improvements the law was intended to promote. By contrast, Aguirre appears to have made the Desert Land Act work as it was intended, eventually acquiring title to 1,120 acres around Aguirre Lake in his own name and that of his daughter, Beatriz.[3] With irrigation, the loamy bottomlands were well suited to agriculture, and Aguirre planted corn, beans, hay, and other crops. The long growing season allowed for two crops to be harvested in good years (*Tucson Post*, 1 January 1903). Aguirre's grandnephew, Yginio, recalled that "as the water receded in [Aguirre] lake they would plant on the dry spots where the water had receded and they would get quite a bit of hay out there and wheat or barley" (Aguirre 1986). It was, in effect, an inverted form of the Indian technique of floodwater farming, employed by the Tohono O'odham and Zuni and adapted by Spanish settlers in New Mexico (Bryan 1929). Aguirre also cut native grasses for sale or stacking as supplemental feed.

Between 1885 and 1888, U.S. Deputy Surveyor George Roskruge worked in the Altar Valley, dividing it into numbered sections, townships, and ranges for disposal under federal land laws. As was typical, settlers were already there; Roskruge's job was to establish the geographical framework that would delineate their claims as well as to note their improvements and occupation of the land. His field notes are by far the most detailed and credible record of the social and environmental conditions of the valley at the time, unequalled for at least thirty years after. From them, we get a glimpse of the vegetation that dominated before widespread cattle grazing.

In 1885 Roskruge surveyed the boundary of the township where the Buenos Ayres headquarters was located. Of the south boundary, two and a half miles south of the ranch, he wrote, "There is fine grass for miles

N[orth] and S[outh] of the line, but no trees and no brush" (Roskruge 1885–86: Book 1494). This is approximately where J. Ross Browne had grown weary of the monotonous grass twenty-one years earlier. The following year Roskruge returned to survey section lines, and his notes describe a well-developed ranch. "The ranch house of Pedro Aguirre . . . [is] substantially built of adobe with shingle roof," he wrote (Book 938: 20). "This township is covered with gently rolling grassy hills and is overrun by cattle and horses. The water used is raised from wells sunk for that purpose or collected in charcos or ponds. Pedro Aguirre has a ranch-house[,] outbuildings and corrals . . . all well built, also a well with steam power to raise water. He is engaged in stock raising" (Book 1512: 163). Of Aguirre Lake he wrote, "This charco covers about 100 acres and is formed by water collected from neighboring lands. . . . In very dry seasons the charco is perfectly dry" (Book 938: 23). He must have seen it dry in the drought year of 1885, for in 1886 it held "a large quantity of water" (Book 937: 81).

Nowhere else in the Altar Valley did Roskruge remark on large numbers of stock; only two other ranches had deep wells, and range conditions were good. At the Warren Ranch (now the Anvil Ranch) near the north end of the valley, one thousand acres of bottomland were fenced; "[a] well was sunk to a depth of 180 feet and an abundant supply of water was obtained, which is raised by steam power" (Book 841). The land was "covered with a luxuriant growth of Grama grasses" (Book 1485), with a scattering of mesquite and palo verde trees. At the Palo Alto Ranch, named for a very tall mesquite tree (Barnes 1960), a man named Russell[4] had a well and steam pump. "The township is covered by a scattering growth of mesquite timber" (Book 1485: 220). In the next township south, a José Espinosa had only a large charco for water, as did a Sacundino [sic]. "There is no living water in this T[ownshi]p [T19S, R9E], water being caught and stored in charcos or water holes" (Book 1485: 147).

At the Poso Bueno Ranch (cf. Granger 1983), later known as Pozo Nuevo, Roskruge wrote the following: "This township contains a fine body of 1st rate land running through the center of the Township from N[orth] to S[outh], which with the aid of water for irrigating would produce large crops. A[ugust] Hemme and others have about sixteen hundred acres of this land under a post and 5 wire fence, having located it under the Desert

Land Act. I am informed that Mr. Hemme intends to conduct water on to this land by means of iron pipe laid from the Arivaca Creek a distance of 10 miles" (Book 897: 100–101). Hemme and his partner, Hubbard Labaree, had fenced seven miles of the loamy bottomlands extending into the next township south, enclosing a total of 2,200 acres.[5]

Water remained the critical limitation on settlement in the valley. Within Hemme's enclosure, a Thomas Taylor had a "small field in which there is some fine corn growing. This season plenty of rain having fallen insured a good crop to those who risked the planting, but as a rule the rains in this vicinity can not be depended on so as to warrant the planting of any large quantity of land" (Book 897: 101–102). Of the next township west, which touched on the foothills of the Baboquivari Mountains at the future location of the Las Delicias Ranch, Roskruge wrote as follows: "The balance of the T[ownshi]p [i.e., that part not in the foothills] is level mesa land covered with fine grasses. Mesquite and Sycamore abound in the arroyos. There is no live water in the T[ownshi]p, but [it] could easily be obtained by sinking [wells] in the arroyos running from the Baboquivari mountains. With water for irrigating, a large portion of this township could be made to yield almost any kind of a crop. This township should be subdivided" (Book 1485).

Roskruge was overly optimistic regarding irrigation, perhaps due to the emphasis being placed on reclamation at the time. Conceivably he may also have overstated the lushness of the grass with an eye to encouraging settlement, though there is no overt indication of this. The preceding year (1885) had been very dry, and if the range had been overstocked, presumably it would have been evident. Yet over and over again Roskruge remarked on townships in the valley: "Grass good. No timber."

The only specific description of grasses in the Altar Valley before the drought of 1891–93 appears in Roskruge's notes on the township surrounding Hemme's enclosure (T19S, R9E): "The bottomland is covered with sacaton and the mesa or uplands with white and black gramma grasses *[sic]*. The timber in the Township consists of [a] scattering growth of mesquite" (Book 897: 102). Giant sacaton *(Sporobolus wrightii)* is a tall bunchgrass adapted to subirrigated and ephemerally inundated floodplains; it produces large quantities of forage, though it is unpalatable to cattle at certain times of year. What Roskruge meant by "white and black gramma grasses" requires careful elaboration. "White" probably referred to *Bouteloua oligostachya* (Toumey 1891; U.S. Secretary of the Interior 1893: 19; Wagoner 1952: 39), now known as *B. gracilis,* or blue grama. "Black" could have meant either *B. eriopoda* (Toumey 1891), its present referent; *Muhlenbergia porteri,* a.k.a. bush muhly (Thornber 1910); or perhaps *Hilaria mutica,* a.k.a. tobosa grass, now known as *Pleuraphis mutica* (Bentley 1898). Chances are that all four grasses were present; all are perennial, warm-season grasses that are excellent forage species.

The incidence of mesquite trees *(Prosopis velutina)* warrants discussion because it played a central role in the twentieth-century history of the valley. We have seen that most accounts emphasize a lack of trees in the uplands. Accounts that describe *no* trees, however, are all from the southern end of the valley. South of the Buenos Ayres headquarters, the pediment zone traverses the valley; this may have favored grasses by depriving mesquite of one of its principal competitive advantages: the ability to tap water from deeper soil layers. Farther north, as Roskruge's field notes make plain, the incidence of mesquites increased. Between Secundino and Palo Alto, he marked individual trees to indicate their township, range, and section and recorded them in his notes. By "scattered" he appears to have meant roughly one or two trees per acre (fig. 5); by his measurements, the trees were four to twelve inches in diameter.[6] Still farther north, densities increased again. In 1871 surveyor S. W. Foreman wrote this about the Avra Valley: "In all parts of the section traversed by this line, there is an abundance of mesquite timber suitable for fuel" (Roskruge 1885–86: Book 1453: 63).[7] In the jargon of government surveying, "timber" was a technical term denoting trees of sufficient quality and quantity to affect the classification or disposal of public lands; by this criterion the Altar Valley indeed had "no timber." It clearly had trees, however, some of which were of significant size.

Much has been made of the need, reported by Pedro Aguirre's descendents, to haul firewood to the Buenos Ayres from Tucson. Photos from 1905 appear to show a complete absence of trees, and this has been construed as the "natural" condition of the valley. Yet the fuel shortage was more likely the product of human harvesting. According to Bahre (1991: 163), a steam engine pumping water from a shallow well consumed a cord of word every ten hours. Griffiths (1904: 35) reported that

Figure 5. Santa Margarita Ranch, northwest of the Buenos Aires Ranch, 1929, with Baboquivari Peak in the background. Note the scattered mesquite trees on the flats beyond the buildings: they are few in number and upright. The corrals were made of mesquite branches laid between pairs of posts. (Courtesy of the Arizona Historical Society, AHS no. 20)

in the Altar Valley, where the wells were very deep, "the fuel used for pumping is almost entirely mesquite from the immediate vicinity." Whatever wood was available—particularly the mesquites in drainages described by Pumpelly, Browne, and Roskruge—must have been harvested in the late 1880s and early 1890s to power the steam pumps at the Buenos Ayres, Palo Alto, and Warren Ranches, with serious impacts on local mesquite abundance. The trees were very likely removed by their roots: The root crown contains a large amount of high-quality fuel, and the roots below it are more easily cut with hand tools than is the main trunk. (This practice is still common in northern Mexico.) Roskruge's map of the Buenos Ayres area in 1886 shows a road to the west-northwest labeled "Wood Road," although nobody named Wood lived in that direction at the time, and Roskruge would have labeled it "road to Wood's," if one can judge from his maps. The obvious inference is that wood was being harvested from the Baboquivari Mountains, where live oak, pine, black walnut, ash, hackberry, and sycamore could all be found. It is reasonable to assume that all sizable mesquites in the central valley were cut first.

This is not to argue that mesquites were as abundant in 1880 as they became in the mid-twentieth century. The evidence clearly suggests that they were concentrated and perhaps confined in depressions and drainages. It is wrong, however, to infer that they were therefore an insignificant part of the landscape. The bottomlands and tributary drainages are unusually numerous and large; even very slight differentials in topography generate striking contrasts in soil moisture and vegetation. The loamy bottomlands extend from the axial drainage up into the lower reaches of the major tributaries. The upland alluvial zone is dissected by scores of roughly parallel drainages, ranging in depth from ten to seventy feet. By way of illustration, consider the fact that Roskruge could stand at the corner of four townships (T18&19S, R8&9E) and easily see flags planted six miles away at corners to the north and east. Yet it was in this very region that he noted the presence of mesquites, some of them a foot in diameter (and, we may infer, at least thirty-five years old and likely twenty feet tall). The explanation lies in the unusual topography. He was standing in the uplands, as were his two flags, on land that, like a draftsman's table, is slightly inclined but otherwise almost perfectly flat. The mesquites grew in the innumerable draws, thus lowered beneath the surveyor's line of sight. Mesquites did not dominate the landscape, and they appear to have been absent in the south end of the valley, but they were by no means absent in the central valley until the need for water prompted early settlers to harvest them.

THE BOOM AND DROUGHTS, 1885–1904

With waters in place, Aguirre and his neighbors could capitalize on the natural abundance of forage in the central Altar Valley. Except for a handful of small homesteads, the valley was entirely public domain, legally open to anyone who sought forage for his stock. Although the precise number of cattle in the valley is difficult to establish, figures for Pima County indicate rapid expansion through the 1880s. According to tax assessor's records, the number of cattle in the county increased from less than twelve thousand in 1880 to eighty thousand in 1884 (even as Cochise County was split off in 1881); the actual numbers may have been double these. Rising prices encouraged expansion, and the recently completed Southern Pacific Railroad facilitated importation of cattle from overstocked and deteriorating ranges in New Mexico, Texas, and the Great Plains. Arizona was a net importer of beef cattle until 1885, when, as would happen again, a market crash coincided with drought. Cattlemen scrambled to liquidate, further depressing prices; those who held out watched the range degrade and their cattle grow thin (Wagoner 1952: 44–45).

The 1885 drought slightly diminished the number of cattle in the county, but above average rains in 1887, 1889, and 1890 reignited the expansion, from 66,500 head in 1886 to 121,377 in 1891. (Again, these are the official tax assessor's figures.) Years of good rain fostered an optimistic determination to hang on (i.e., not sell) in dry years. Arizona's territorial governors served as boosters, tirelessly promoting the cattle boom in their reports to the U.S. Secretary of the Interior. They portrayed it simultaneously as the natural result of biogeography and as a relative economic advantage over other areas. "In Arizona a day without bright sunshine is so rare as to be remarkable, and every month in the year cattle run on their ranges and find no lack of feed. These favorable climatic conditions make Arizona the stock-raisers' paradise. . . . [T]his Territory can produce beef more cheaply than any grazing region in the United States" (U.S. Secretary of the Interior 1885: 6–7). The dependence of cattle production on nature was differentially acknowledged: Prosperity was constructed as a "natural" result of rainfall and grasses, whereas low prices and recession were viewed as anomalous social products born of panic during drought. Making money from nature's work was naturalized and thereby effaced, whereas losing money was viewed as an unnatural anomaly (cf. Smith 1978: 187–188).

The rains failed again in 1891. Measured at Tucson, precipitation for the year was only 63 percent of the annual mean (fig. 6). In September Professor J. W. Toumey of the Arizona Agricultural Experiment Station published a short bulletin, warning of a vicious cycle already in motion. "[W]here drought and overstocking both combine, and the grass that

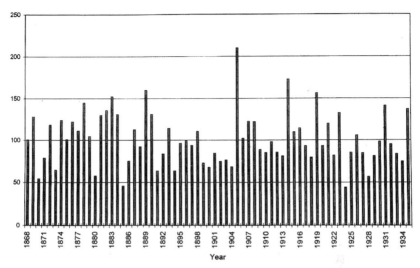

Figure 6. Percentage of average annual precipitation for Tucson, 1868–1935. Note above average rains every year but one from 1874 to 1884, when herds rapidly grew in size, and below average rains from 1899 to 1904, followed by the deluge of 1905. (Reproduced from Bahre and Shelton 1996, with permission)

does not burn out from the effects of the hot sun, is continually eaten close to the ground by hungry cattle, the range is in poor condition to produce feed for the following season. The repetition of this process year after year cannot help but decrease the supply of grasses on the range" (1891: 7). He went on to note: "There is a limit to which any range can be profitably stocked. If we go beyond this limit it will not only be a detriment to the permanency of the range but will be detrimental to the stock as well" (1891: 8). It was a prescient warning, but it went unheeded. Most ranchers attempted to hold out, prices for cattle being very low at the time; cattle in the county decreased only slightly, from 121,377 to 116,604 in 1892. Continued dry conditions in 1892 cut this figure by more than half, to 49,599. The rains did not return until mid-July 1893. Southern Arizona was the hardest hit region in the territory; 50 to 75 percent of cattle perished on the range (U.S. Secretary of the Interior 1893: 21, 1896: 20–24). One Arivaca rancher later testified that he lost six hundred cattle in the drought, including some that perished in the ciénega, too weak to extricate themselves from the mud after venturing in for water.

With the benefit of hindsight, it can be inferred that the drought of 1891–93 pushed most of southeastern Arizona's ranges across an ecological threshold. "Allowing cattle to starve during droughts on southeastern Arizona's overstocked ranges probably contributed more than any other factor to rangeland deterioration at the turn of the twentieth century" (Bahre and Shelton 1996: 1). The damage done can hardly be overstated. In photographs from the time, "hundreds of square miles of rangeland are denuded of cover; the grasses, even the sacaton in the bottomlands, are grazed to the ground; the hills are covered with cattle trails; erosion is rampant; and the oaks and other trees have browse lines" (Bahre 1991: 113). David Griffiths, a government botanist at the time, whose research I will consider in detail in chapter 3, described Arizona as "a region more completely divested of range grasses than any other in the entire country" (1901: 23). Large areas where grass had grown waist high were reduced to dirt. It is less clear, however, how much impact the great drought had on rangelands in the Altar Valley.

It appears that Aguirre stocked the Buenos Ayres at rates six to ten times greater than those prevailing today. The *Arizona Daily Star* reported "thousands of cattle on his ranch" in 1886 (7 May). Five years later, when the first great drought was just beginning, the paper reported that he owned five thousand head of cattle (2 June 1891). Aguirre's grandnephew confirms this number and adds seven thousand sheep to it (Aguirre 1986). These figures, however, are very rough because even Aguirre did not know precisely how many animals he owned. The estimate of five thousand head was derived, according to his grandnephew, by counting the pieces of ear cut from calves at roundup and multiplying by five (Aguirre 1975: 271). It is impossible to know the precise extent of the Buenos Ayres's range at the time because it was unfenced, but it appears to have extended from near the Mexican border on the south to around Arivaca Road or Secundino on the north. By informal range rights, it comprised at least thirty thousand acres and perhaps as much as fifty thousand, suggesting that Aguirre stocked cattle at one head per six to ten acres; including the sheep, the stocking rate was roughly 80 to 130 animal units per section. Today the norm for the area is ten to twelve.

Aguirre's financial circumstances were tenuous throughout the late 1880s, even as his herds grew. It appears that most of his animals were pastured on contract for others or bought on credit. In 1884 Aguirre terminated partnership with a John Doggett of Kansas (*Arizona Daily Star,* 15 October 1884). We do not know the substance of this arrangement or the reasons for its demise, but it suggests that Aguirre continued his earlier practice of forming partnerships with cattle and sheep owners from elsewhere. We also know that he frequently imported cattle from Mexico for fattening on the Buenos Ayres. The abrupt drop in cattle prices in 1885—-from more than thirty dollars to less than ten dollars a head (Wagoner 1952: 45)—appears to have caught him overextended. In 1886 *Hoof and Horn* (4 August) reported that he and Juan Elias had negotiated with "Henry Amey and other well known New York capitalists" to sell their cattle, ranges, and water rights. "These ranges cover an area of forty-five by seventy miles and extend far into the State of Sonora, Mexico" (BANWR archives). Aguirre was to remain as manager, and twenty thousand cattle were to be stocked. The deal never went through, perhaps owing to the cloudy title of Elias's large land grant straddling the border at Nogales. It is suggestive, nevertheless, of circumstances at the time: Outside capitalists were interested in investing in ranches, while ranchers themselves were buffeted by shifts in the weather and the markets. According to his grandnephew, Aguirre was bankrupt twice, once in 1891, when the *Arizona Daily Star* (2 June) reported that his creditors were dividing up his herd among themselves. (The figure of five thousand head may have been crafted for these creditors.) Chances are that his debts stemmed from ranching and mining; the only evidence we have is fragmentary legal correspondence relating to money he owed to the Arivaca Land and Cattle Company (UASC, Selim M. Franklin papers). It is reasonable to assume that Aguirre's financial problems led him to stock the ranch even further than he might have otherwise in an attempt to cover his obligations.

The great drought of 1891–93 appears to have spared much of the rest of the Altar Valley, however, because of the relative lack of water sources. The number of wells in the central valley was small—three or four at most—and tanks and charcos undoubtedly dried up. Given that cattle rarely graze more than three miles from water, it is likely that a substantial amount of the Altar Valley remained nearly ungrazed, in contrast to valleys such as the San Pedro and the Santa Cruz, where water was more evenly distributed. This supposition is supported by the only solid evidence available: photos taken in 1893 of two boundary monuments in the valley, located .2 and 2.5 miles from water, respectively. Almost no ground cover is visible around the first monument, whereas a thin but unmistakable cover of grass can be seen surrounding the second (Humphrey 1987: 293–295).

The greatest effect of the 1891–93 drought in the valley may have been the impetus it provided for the further construction of wells. Two were dug at Poso Bueno Ranch by June 1891 (Granger 1983: 495). The deep well at Secundino was drilled by 1912 (Barnes 1960). When Kirk Bryan surveyed water sources in 1917, he found twenty-two wells between the Anvil (formerly Warren) Ranch on the north and the La Osa Ranch on the south, seven of which were in the central alluvial area. He published these remarks in 1925: "The grassy plains of the Altar Valley are much more suitable for cattle raising than other valleys of the Papago country. Success was assured for the stock business when it was found possible to develop water either in the edge of the mountains or by deep wells in the center of the valley. The chief watering places have been in existence for more than 30 years" (Bryan 1925: 375).

Although rains were below average again in 1894, they were reasonably good for the following four years; assessor's cattle figures for the county increased to around sixty thousand. Then another, longer drought hit from 1898 to 1904, and cattle again perished on the range in large numbers. At the end of this second great drought, the cattle population of Pima County was less than twenty thousand by official figures: one-sixth of its 1891 peak. The combined effects of the two droughts are documented in a number of photographs taken in or near the Altar Valley.

David Griffiths traveled through Arivaca to Sasabe in March 1903, returning to Tucson via Robles Junction at the north end of the valley. He made another trip to Arivaca the following month. Several of his photographs were published in a U.S. Department of Agriculture (USDA) report the following year; others were published in 1996 (Bahre and Shelton 1996). One, taken at the La Osa Ranch immediately southwest of the Buenos Ayres, shows a windmill surrounded by corrals and buildings in

Figure 7. Headquarters of the La Osa Ranch, southwest of the Buenos Ayres Ranch, 1904. Heavy grazing and prolonged drought had reduced grasses to insignificance; only unpalatable shrubs, a few mesquites, and bare ground are visible. (From Griffiths 1904)

the background; in the foreground, only burroweed and a few leafless mesquite trees grow from rocky soil (fig. 7). Two photos from Robles Junction depict dead horses and a pile of bones; the soil is exposed, and plant life is almost completely absent. A photo taken "near Arivaca" reveals the bones of thirteen cattle in an arroyo; the adjacent hill is stripped of grass, and rocks have been loosened from an eroding cattle trail.

ENTRENCHMENT AND THE FORMATION OF THE ALTAR WASH

Griffiths's 1903 photo taken near Arivaca reveals something else as well: a vertical arroyo bank in the background, roughly six feet high. This is the earliest documentation of downcutting in the area. Arroyo formation in the Altar Valley was systematically studied by Cooke and Reeves (1976), who concluded that it began after 1886 and before 1923. They were unable to specify the date any further because of lack of documentary evidence. They and other researchers have long remarked on the incidence of entrenchment throughout the Southwest between 1860 and 1900. In southern Arizona, the period 1885–95 encompasses major downcutting episodes for the Santa Cruz, San Pedro, and San Simon Valleys; hundreds of acres of the Santa Cruz floodplain washed away in August 1890 alone (Bryan 1922; Dobyns 1981; Hastings 1959; Hastings and Turner 1965). Scholarly debates have centered around whether "natural" or human causes were responsible. Overgrazing, climate change, irrigation diversions, railroad embankments, fire suppression, the extermination of beavers, timber cutting, and even earthquakes have been cited as contributing factors. All accounts agree that the cattle boom must have contributed to arroyo formation by diminishing vegetation cover and thus increasing rates of runoff. Most also conclude, however, that grazing was not the sole cause because downcutting appears to have been triggered by other factors, such as irrigation diversions cut into stream channels. There is also convincing evidence of prehistoric episodes of downcutting followed by periods of aggradation (Bryan 1940). I present the evidence for the Altar Valley here without attempting to resolve this larger debate.

It is well established, from Roskruge's field notes and other accounts, that in 1886 the loamy bottomlands of the Altar Valley were thickly vegetated with sacaton grass and uninterrupted by any entrenched interior drainage. Rancher Manuel King recalled that before 1900 he could ride a horse ahead of any flood because the water moved slowly in a broad sheet over the flat plain (SCS 1992). When he built his corrals at the Anvil Ranch shortly after 1908, he put the bottom boards a foot off the ground to allow the floodwaters to pass underneath. Sheet flooding of this type in the Santa Cruz Valley was described by W. J. McGee (1897: 100): "The front of the flood was commonly a low, lobate wall of water 6 to 12 inches high . . . and it was evident that most of the water first touching the earth as the wave advanced was immediately absorbed and as quickly replaced by the on-coming torrent rushing over previously wetted ground. . . . Such were the conspicuous features of the sheet flood—a thick film of muddy slime rolling viscously over a gently-sloping plain." The main north-south road in the Altar Valley was originally located in the floodplain to avoid the rougher terrain of the adjacent mesa lands.

Griffiths's photo corroborates testimony presented in a 1908 trial before the U.S. Land Office, which suggested that entrenchment began on

Arivaca Creek between the two great droughts. Through the first drought, the ciénega remained a muddy marshland, with the springs providing moisture. By the end of the second drought, a deep trench had formed, draining the ciénega and causing the water table to drop throughout the floodplain. The trial hinged on whether this entrenched land should be considered "desert" or not. If it could be "remuneratively cultivated" without irrigation, then it could not be "desert" under the federal laws. Floodplain farming, including the cutting of natural hay in bottomlands (which had been done in the past), did not fit comfortably into this legal framework; moreover, it was unclear if such farming could be practiced any longer. S. F. Noon, lawyer for the plaintiff, cross-examined Dr. Joseph Ball, an Arivaca resident and witness for the defense, on this point:

Q (Noon): That is the water spreads out, over the whole valley, there is no deep channel?
A (Ball): There is a channel now in places that carries the water and the land that used to be overflowed has become desert land.
Q: A big deep channel?
A: Yes sir.
Q: And the water runs in the channel?
A: Yes sir, and not only that but it drains the water out of the ground.[8]

According to Bryan (1922: 74), Arivaca Creek ran in a trench fifteen feet deep and about two hundred yards wide near the town site in 1917.

Evidence for the central Altar Valley is far from conclusive, but it suggests that the arroyo there formed slightly later than the one along Arivaca Creek. The earliest documentation of the Altar Wash (sometimes called the Brawley Wash) comes from livestock inspector Willard Wright, who testified in the same 1908 trial. Asked about the relation of sacaton grass to flooding, Wright remarked that on "Baboquivari flat . . . [t]he sacaton will be up above on the mesa and the water running down below in the washes." In context, Wright appears to mean the area around the Palo Alto and Pozo Nuevo Ranches, where Hemme had built his enclosure in the late 1880s. Cooke and Reeves (1976: 56ff.) cite the enclosure as one likely point of initial entrenchment, along with other "drainage concentration features" documented by Roskruge that correspond with the subsequent Altar channel: Warren's enclosure farther north, two stretches of the main north-south wagon road, and eight charcos. Taken together, these features formed a nearly continuous path of soil disturbance approximately twenty-seven miles long down the center of the valley. Cooke and Reeves hypothesize that these features determined the location of entrenchment; they remain more agnostic about ultimate causes.

To Cooke and Reeves's argument, we can add several observations. First, in 1908 the trench was far enough from the Anvil Ranch that Manual King built his corrals as though they would not be affected by it soon. Second, by that time there was trenching farther south that was continuous enough to strike Willard Wright as a wash (or washes), not simply as a charco. Third, these facts suggest that the Altar Wash began at the upper end of the watershed and moved downstream over time, the opposite of arroyos such as the one on the Santa Cruz River, which grew by headcutting upstream. Finally, Arivaca Creek experienced downcutting before the main Altar Valley drainage did, and the trench there remained wider and deeper than the Altar Wash into the 1920s. This would correlate generally with a higher degree of human impact—grazing, farming, irrigation, hay cutting, and so forth—in the Arivaca area during this period.

These items would scarcely be worth mentioning if they did not dovetail with a hypothesis that Cooke and Reeves did not consider. They acknowledge the need for "a concentrated supply of water" to initiate entrenchment, and they speculate that Hemme's waterline, if it was built, might have been the source (1976: 57). There is no solid evidence, however, that it was built at this time, and it is hard to imagine a pipe at most a couple of inches in diameter resulting in twenty miles of entrenchment. Absent from their analysis is any consideration of Aguirre Lake, however. Might the dam have burst, sending downstream the massive pulse of water necessary to link the charcos, the two stretches of road, and the enclosures by excavating a continuous arroyo?

There is indirect evidence to support this hypothesis. We have seen that grazing during the 1898–1904 drought denuded large areas of the

Altar Valley, which would have dramatically increased runoff rates. Measured at Tucson, rainfall in December 1904 was almost an inch; in January 1905, 2.25 inches fell. The *Arizona Daily Star* of 29 January included this item: "Pedro Aguirre is in from his Buenos Ayres Ranch. Don Pedro says that he has two hundred acres in barley and will have a large acreage in beans. *His dam is full to overflowing,* solving his water problem for the season" (emphasis added). In February, however, 4.15 inches of rain fell, followed by another 3.88 inches in March: more than twice the amount for that two-month period than in any other year between 1868 and 1921 (Bryan 1925: 40). If just over three inches had filled the dam "to overflowing," then an additional eight inches presumably resulted in severe flooding. (The Altar Valley typically receives more rain than Tucson, so these figures should probably be adjusted upward). Bryan (1925: 145–150) reported that either no spillway or only an inadequate one existed at Aguirre Lake prior to 1915; in either case, a collapsing spillway would have had much the same potential energy as a breached dam. The floods of winter 1905 are legendary. They took out bridges and railroad tracks throughout southern Arizona (Dobyns 1981) and partially rerouted the Colorado River into the Salton Sea (DeBuys and Meyers 1999). In Pima County, the floods extended all the way to the Santa Cruz flats northwest of Tucson and did damage elsewhere in the Santa Cruz watershed.[9] February and March 1905 must be considered the most likely date of the creation of the Altar Wash, and Aguirre Lake its proximate cause. An almost identical incident would occur in 1917 (see next chapter).

Over the course of the twentieth century, the Altar Wash grew from a trench some six feet deep and six feet wide to its present size: fifteen to twenty feet deep and up to fourteen hundred feet wide in places. By the late 1930s it extended from Aguirre Lake north to the Avra Valley, a distance of more than forty miles (Andrews 1937: 167, 173; fig. 8).[10] Analysis of aerial photographs by the SCS (1992: 11) concluded that the average width of the channel increased from 217 feet in 1936 to 410 feet in 1987. Arroyo formation not only removes bottomland soil by erosion, but also strongly affects vegetation in the adjacent bottomlands by lowering the alluvial water table (Bryan 1928; Hendrickson and Minckley 1984). The former floodplain now remains dry even in very large floods,

Figure 8. Entrenchment of the Altar Wash circa 1929. Before 1905, there was no trench or defined channel in the axial floodplain of the Altar Valley. The soil lost to the arroyo was among the valley's most productive. By changing runoff and infiltration patterns, the Altar Wash dramatically impacts vegetation on the terrace above. (From Brown 1931, p. 22a; courtesy of the Arizona Historical Society)

while the waters rush by in the entrenched arroyo. For decades owners of the Buenos Aires Ranch would devote considerable resources to mitigating the effects of the Altar Wash.

The Altar Wash did not attract much attention at first, however, and it does not appear to have greatly affected the ranch during Aguirre's lifetime. On New Year's Day 1903, the *Tucson Post* reported that "Mr. Aguirre has one of the best ranches in the Territory. He has one lake half a mile wide and three quarters of a mile long that holds water enough to irrigate several hundred acres of land. Two crops a year of corn and beans and such other stuff are raised on the place." Aguirre also cut native grasses for sale or stacking as supplemental feed; later owners supplemented this practice by sowing Johnson grass in the bottomlands, largely displacing sacaton.

Aguirre passed away in Tucson in 1907. His estate sold the Buenos Ayres Ranch in 1909 to William Coberly's La Osa Cattle Company for six thousand dollars. Descendents of Aguirre's brothers are still ranching today, northwest of Tucson at Red Rock (cf. Aguirre 1969).

CONCLUSION

In his book *Hunters, Pastoralists, and Ranchers,* social anthropologist Tim Ingold (1980) isolates two major components in the transformation from pastoralism to ranching. One relates to the source of production and the other to the control of land. He does not examine the cattle boom in depth, but his arguments provide a powerful theoretical framework for examining the history and ecology of ranching in the American West. By Ingold's model, the cattle boom combined features from both pastoralism and ranching.

Ingold distinguishes between "two spirals of accumulation, one distinctively pastoral and based on the natural reproduction of herds, the other distinctively capitalist and based on the exchange of products, through the medium of money, for factors of production including labour and animals" (1980: 3). In a pastoralist economy, nature produces the surplus in the form of live animals, which the pastoralist simply protects and then appropriates. "The economically productive work on which herd growth ultimately depends is performed by the animals themselves, and not by their human guardians" (Ingold 1980: 222). In ranching, by contrast, surpluses take the form of money and are the product of combining various inputs obtained in the marketplace. Natural productivity gives way to value produced by labor and congealed in the inputs and outputs. This shift corresponded with the cattle boom itself, expanding across the West as the railroad and cattle trails connected frontier rangelands to eastern markets. As early as the 1870s cattle producers were buying (as well as selling) their animals in the marketplace.[11] In this regard, the historians' conventional definition of ranching is correct.

The second condition for ranching did not develop until later, however. It concerns access to land. Both pastoralism and ranching involve divided access to animals (i.e., animals are the exclusive property of individuals or households, in contrast to hunting, in which animals are common property until killed). In pastoral economies, however, access to land is common, whereas in ranching access to land is divided by ownership or use rights (Ingold 1980: 4–5). By this criterion, the transformation to ranching occurred between 1907 and 1934, when exclusive use of public lands for grazing was institutionalized in the form of leases.[12] Before leases, under the "open range" system, access to land was legally shared (though individuals made valiant and sometimes violent efforts to monopolize portions of the range).

Understood in these terms, ranching is not defined by cattle or capitalism alone, but by the social relations surrounding livestock and land. It did not simply spread across North America with the cattle boom, but formed during the seven decades between the Civil War and the Dust Bowl. Spanish and Mexican cattle raising was a form of pastoralism, and the Anglo-American open range was a transition period, a hybrid of pastoralism and ranching (cf. Ingold 1980: 240–241). It is no coincidence that this transition produced by far the worst ecological damage ever done to western rangelands because the hybrid combined the most destructive tendencies of both forms of production. Credit and the railroad enabled rapid expansion of herds, and common lands gave each individual the pastoralist's incentive to overstock (Ingold 1980: 207). Production became market mediated and profit driven before land tenure arrangements were established that could effectively limit the number of animals grazing the land.

The cattle boom imposed spatial and temporal assumptions that were incompatible with natural range productivity, especially in arid and semi-arid regions such as southern Arizona. As a natural product, forage grasses depended on local and regional ecological conditions for their reproduction. These conditions were highly variable and subject to periods of drought, which could dramatically reduce the amount of feed available for stock. As an input in the accumulation of capital, however, the grasses were a link in a much larger financial system that demanded regular, reliable returns, regardless of interannual variations in forage. The spatial distance between investors and cattle herds was bridged by annual reports, telegrams, letters, and financial arrangements—banks, stock markets, corporations, and other instruments—that used paper shares or notes to represent the value of animals. Reports could be fudged, of course, and some stateside managers used the natural disasters of the 1880s as opportunities to revise their paper herds closer to reality. Furthermore, the notes and shares were tradable goods in their own right, with a market value that could move up or down regardless of the cattle, the range, or the weather. The real "killings" were made here, in buying shares low and selling them high. The *Economist* and other financial journals knew quite well that the boom was at least potentially a confidence game, and accurate information about the material basis of shares was highly sought after (Frink et al. 1956; Graham 1960; Gressley 1966).

The elastic relation between paper and reality was the structural keystone of the system, a source of great potential and equally great risk. The cattlemen needed capital. They got it from faraway investors, who sought dividends and rising share prices and who could not see the effects of their investments. In Great Britain as well as on the frontier, cattle were promoted as a way to achieve wealth without work. Ideally, reality and paper mirrored one another, the investors' pastoralist fantasy being one of "natural" increase through capital appreciation. Volatile markets, unpredictable weather, and indeterminate land tenure, however, defied this ideal. The result was overstocking, severe overgrazing during droughts, a host of related environmental problems, and cattle dying on the range.

In summary, four conditions made the cattle boom possible. First, a natural process was subsumed within market-mediated processes of capital accumulation. Cattle eating free grass, drinking readily available water, and generally following their natural instincts produced more cattle; shipped to markets elsewhere, they could be converted, for a time, into huge profits. Second, this form of capital accumulation was more profitable on the open range of the recently pacified West than elsewhere, and it held out the possibility of greater profits than other forms of production and accumulation. A competitive advantage was found and exploited in the natural geography relative to the rates of return prevailing in national and international financial markets (White 1991: 243). Third, means were available for the large-scale movement of capital (as money and as cattle) across this economic-geographical boundary. Fourth, a huge surplus of finance capital existed and was looking for a form and place to undergo transformation and reproduction as capital. Its availability enabled ranchers to expand their herds immensely, while its demands compelled them to disregard ecological limits until it was too late. If any of these conditions had not been met, the cattle boom and subsequent ecological damage to western rangelands would not have occurred, at least not on the scale that they did. It was not cattle, or the railroad, or culture that stripped bare the grasslands of the arid and semiarid plains, but all three, driven and united by an unprecedented *scale* and *rate* of capital immigration.

In the Altar Valley, the cattle boom and the droughts of 1885–1904 were mutually reinforcing in three respects. First, the boom created the social and economic conditions for overstocking. Aguirre, by all the evidence, ran livestock on credit or in partnerships, which encouraged rapid stocking increases in good times and discouraged destocking when drought caused cattle prices to drop. This was typical of the cattle industry at the time. Second, nature was modified to accommodate cattle raising. Artificial waters, necessary to take advantage of the abundant grasses of the Altar Valley, were further motivated by the droughts. Those who could not afford wells were "drouthed out" and compelled to sell to those who could. If the 1891–93 drought spared large stretches of the valley, as I have suggested it did, later droughts did not, precisely because investments were made to provide water where there had been none. Third, the two great droughts undoubtedly had a greater effect on vegetation than either one would have had by itself. Stocking rates countywide were lower during the second drought, but they may have been the same or even higher in the Altar Valley. Moreover, if the range had not yet recovered from the first drought, the damage from the second may well have been greater and longer lasting even with smaller numbers of cattle.

3

The Formation of Ranching

Eighteen ninety-three marked the end of the cattle boom in Arizona for two reasons. The first was local and ecological: Overgrazing during the drought of 1891–93 dramatically altered the native grasslands and diminished the number of stock the land could support, achieving in Arizona what overstocking, blizzards, and droughts had achieved earlier across the Great Plains. Unlike in the 1880s, moreover, in the 1890s there was no longer any free "virgin land" available. At the very least, investments in artificial waters were necessary to utilize the grass that remained. The second reason was international and economic: Panicked by the depression of 1893, "investors suddenly turned off the flood of dollars, with no warning," initiating a "money drought" throughout the American West (Worster 1985: 131). Many large corporate land and cattle companies liquidated and were replaced by smaller outfits, usually owned and operated by individuals or families (White 1991: 270–271).

Those who emerged from the devaluation and restructuring thus faced two problems at once: the degraded condition of the land on which their herds depended and severely contracted access to credit. These two problems would dominate the work of ranchers and government agencies as they endeavored to complete the formation of ranching over the decades to come. The ecological problem was openly expressed and became the object of significant scientific research, which eventually evolved into the discipline of range science. The economic one remained largely unstated, the private affair of ranchers and their financial backers. The two were nonetheless closely related. Public land managers and scientists always focused on "the natural resources," but underlying this focus, structuring the very way they approached resources, was the need to stabilize conditions for capital investment in the cattle industry after the collapse of the boom.

For reasons described in the previous chapter, events in the Altar

Valley lagged behind the regional pattern. Hemme and Labaree's Palo Alto Ranch succumbed in 1892, paying off its employees with cattle. (One employee was Manuel King, who used the severance to start his own herd. His descendents continue to ranch in the valley today.) But most of the valley's major operations survived into the next drought. In 1902 Colonel William Sturges "of Chicago board of trade fame" (*Arizona Daily Star*, 1 January 1937), who had homesteaded the Las Moras Ranch and later acquired the La Osa and Palo Alto Ranches, sold his holdings to William Coberly. Coberly organized all the ranches into the La Osa Cattle Company. In 1909 he bought the Buenos Ayres Ranch from Aguirre's estate, thereby securing control of most of the southern two-thirds of the Altar Valley. After statehood in 1912, Coberly initiated the process of securing leases on selected state lands in the valley (see below). Other than this consolidation, his most lasting mark on the ranch was to normalize the spelling of its name, changing it to Buenos Aires.[1]

Between 1913 and 1915, La Osa passed out of Coberly's control—apparently in settlement of debts—into the hands of Jack Kinney (fig. 9), who reorganized it as the La Osa Live Stock and Loan Company.[2] Kinney typified the large-scale cattlemen (many from Texas) who expanded into southern Arizona in the second decade of the century (Wagoner 1952: 57). He was born in Dixon, Illinois, in 1872. He moved first to Texas and then to Montana in 1890, where he became involved in banking and ranching. He prospered and rose to some prominence, serving eight years as a state senator. With strong connections to banks in Texas and the Midwest, Kinney first came to Arizona in 1913 to broker loans and contract for cattle to ship to ranches in Montana, North Dakota, and California. He lent more than $1 million to Arizona cattlemen before moving to the state in 1915. Later he became active in the Arizona Republican Party, serving as a county supervisor from 1928 to 1932 and as candidate for governor in 1932. He also founded Tucson's annual rodeo, "La Fiesta de los Vaqueros," which to this day is a local holiday and festival.

Completing the formation of ranching required, at a minimum, that the open range be divided into bounded parcels for individual use. As early as 1893 the territorial governor had asked the U.S. Secretary of the Interior for authority to issue leases on the public domain, casting it as a means of restoring past prosperity: "The ranges would not be overstocked,

Figure 9. Jack Kinney, president of the La Osa Live Stock and Loan Company. Kinney facilitated an influx of capital into southern Arizona ranching after the collapse of the cattle boom. (Courtesy of the Arizona Historical Society, AHS no. 47043)

and cattle would be fat at all times of the year, as they were in the early days of the range industry in the Territory" (U.S. Secretary of the Interior 1895: 40; see also 1893). U.S. Forest Service (USFS) lands in southern Arizona gradually came under leases after 1907, but these did not encompass the grassland valleys where grazing was most concentrated. In the Altar Valley, leases were instituted shortly after statehood in 1912 by a series of steps that converted the open range first into State Trust lands and then into grazing allotments for exclusive use by lessees. La Osa pursued this matter aggressively, securing nearly two hundred thousand acres of state leases by 1920 to complement its twenty thousand acres of deeded land.

The formation of ranching was not simply a matter of leases, however. The crisis produced by the boom provoked a larger response from government to fix the damaged land in partnership with ranchers. The public face of this effort was scientific, but it is better understood in terms of Ingold's (1980) other criterion for ranching: the subsumption of natural productivity by market processes. In the aftermath of the boom, the list of productive components for which ranchers had to pay money grew to include forage (in the form of leases), cattle, fences, water (in the form of wells and tanks), supplemental feed, and a variety of vaccines and medicines. Accordingly, the capital invested in ranching increasingly took the form of "improvements" aimed at restoring the range's preboom productivity, which was understood as a "natural," and therefore timeless, attribute of the land. La Osa invested large sums of capital in the Buenos Aires Ranch, but the outcomes were decidedly mixed. Profits were high during World War I, but the company's debt-to-asset ratio escalated rapidly during the droughts and economic downturn that followed the war. Range conditions rebounded from 1904, but as annual grasses replaced perennials, the variability of forage production with rainfall became only more pronounced. The drought of 1920–21 resulted in another episode of cattle dying on the range.

Government science played a key mediating role in range reform. Researchers endeavored not just to understand natural processes but also to find economical ways of rendering them tractable for private ranchers.[3] The lexicon of these efforts was scientific knowledge generated by government researchers; the mechanisms were land management bureaucracies and agricultural extension programs. Leases, it was said, would bring stability—for soils, the cattle industry, and rural communities—whereas research and improvements would bring progress, measured in productivity and profits. Before turning to the La Osa Live Stock and Loan Company, I will briefly examine the origins of government range research.

THE DIVISION OF AGROSTOLOGY

In 1895, suddenly alarmed by the destruction of western rangelands, Congress authorized fifteen thousand dollars for grass and forage plant investigations to be undertaken by a new Division of Agrostology within the USDA (Chapline 1944: 129). On first glance, the division appears to have been little known and inconsequential. Within eighteen months its name was being translated in brackets in official publications ("[Grass and Forage Plant Investigations]"), suggesting that people did not know the meaning of "agrostology." (The term appears in few dictionaries today.) In 1901 it was merged with the Division of Botany to form the Bureau of Plant Industry. Yet in its brief six years, the Division of Agrostology set the course of federal range research and policy for decades to come. Its publications reveal that a more adequate understanding of rangeland ecology might have emerged sooner than it did had economic imperatives not intervened.

The division's first circular portrayed its mission in terms that were simultaneously patriotic, biological, and economic. Acting Agrostologist Jared Smith proclaimed that grasses were "the most important family of plants" and that American grasses were "one of the richest legacies of our country." They enabled the United States to export millions of dollars worth of meat products instead of paying to import them. Yet this endowment was, he wrote, being squandered across the West and the South. "Our native grasses and clovers are being driven out by foreign species, not through any superiority of digestibility or chemical composition, but because we have not found the time to discover whether our own species are not far better suited for our soil and climatic conditions" (Smith 1895: 2).

The purpose of the Division of Agrostology, Smith argued, was to perform the work elsewhere done by generations of farmers: to collect and test forage plants in order to select and domesticate those best suited to American conditions. This had always been a slow, small-scale, trial-and-error process, but in government hands it would be done rationally and scientifically so that results might be achieved more quickly and applied more broadly. The goal was maximum national production, measured in time and money. "We want to know what plant will provide the greatest amount of the most nutritious forage in the shortest season at the least expense to the farmer. . . . In short, we want the best, and we believe the best can be grown on American soil from native species" (Smith 1895: 3).

Smith's appeal to botanical nativism was more rhetorical than scientific, and it disguised two fateful contradictions. On rangelands, there was little reason to presume that native plants would be more economical than "foreign species" when *cultivated* for forage, especially if intensive selection and hybridization were necessary first. The native grasses, after all, had evolved without cultivation. By collapsing the distinction between wild and domesticated species, Smith implicitly grouped rangelands in the same category with planted pastures. Since its creation in 1862, moreover, the USDA had invested heavily in collecting plants from around the world, seeking profitable cultivars for U.S. agriculture (Kloppenburg 1988). For most of the twentieth century, nativeness would remain subordinate to the quest for economic efficiency.

These contradictions were perfectly captured in the division's first project: establishing two "experimental grass gardens," one at USDA headquarters in Washington, D.C., and a larger one in Knoxville, Tennessee. American farmers and ranchers were encouraged to send herbarium specimens and seeds to the division, while USDA agents delivered grass and forage species from around the world. All were planted in the gardens, where they could be carefully studied and cultivated. Agrostologist F. Lamson-Scribner (USDA 1898: 164) likened each species to "an interesting chapter in a book thus made complete" by being assembled together in the gardens. The analogy is apt because the division's activities amounted to a massive project of official documentation: collecting, identifying, studying, and judging botanical phenomena. The gardens served not only as areas for observation and study, but also as sources of seeds for farmers and ranchers to test in the field. Apparently, no one asked the prior question: Could a carefully tended garden accurately represent natural range conditions? Or, put the other way, could anyone really envision cultivating millions of acres of range?

The idea that native forage species could be restored by scientific ingenuity and human manipulation also shaped the earliest fieldwork of the division, undertaken by order of the Secretary of Agriculture in 1897. Jared Smith and H. L. Bentley secured the use of two sections of land in Texas to study "the native grasses and forage plants, their abundance and value, their preservation, and the possible methods to be employed in restoring those ranges that have become nearly valueless through overstocking or other causes" (USDA 1898: 170). Smith worked near Channing in the Texas Panhandle while Bentley conducted experiments outside of Abilene.

The experimental designs were nearly identical at both locations. The land was divided and fenced into pastures. Some pastures received mechanical treatments: harrowing, disking, broadcast seeding, or furrowing to capture runoff and wind-borne seeds. On all pastures, periods of grazing alternated with periods of rest. Smaller areas were set aside for garden experiments to identify species and techniques for cultivating pasture. Bentley, working in an area of higher rainfall (thirty inches per year on average), concluded after three years that restoration was both possible and economical, even on degraded ranges. "The greatest successes were secured with the native grasses," Bentley observed (1902: 21), and the costs of fencing and treatment were amply repaid in increased forage production. (Although the treated areas responded more than the untreated ones, Bentley also noted that the alternation of grazing and rest periods contributed to recovery in all the pastures by allowing grasses to mature seed [1902: 34]. This insight would be lost, however, in much subsequent research.) Bentley confidently asserted that three years' work along the lines described would yield similar results on rangelands elsewhere in the region. He arrived at the time period by observing that "two comparatively unfavorable seasons [i.e., drought years] are very likely to be followed by one at least that is specially favorable" (1902: 27). This was, in fact, what had happened during his research.

The central concept of Smith and Bentley's work was carrying capacity, which Smith (1899: 10) defined as "the number of stock which may be supported upon [any large area] during its poorest years." Where they got the concept is unclear, but the benchmark for their analysis was the "original" productivity of the land, measured in acres per animal or animals per square mile. From a questionnaire completed by cattlemen in eighty-two Texas counties, Smith concluded that the carrying capacity of Texas rangelands had declined 40 percent since 1874, "a very large loss in what may be called the capital value of the grazing lands within a very short period" (1899: 13). Areas in Arizona and New Mexico, he reported, could support only ten cattle per section, "and the land there is almost bare of vegetation" (1899: 9).

Carrying capacity contained the same contradiction as the grass gardens. On the one hand, it was a natural attribute, intrinsic to a landscape: "original" productivity was ahistorical, at once the benchmark and the goal for restoration efforts. On the other hand, it could be manipulated and supplemented by human efforts and scientific improvement. Bentley conceded that his experiments had failed to restore the land near Abilene to its "original capacity," but the carrying capacity had more than doubled, and three more years' work would double it again, he asserted. Both he and Smith recommended investments in artificial water sources, cultivated hay crops (using irrigation if necessary), and silage systems to preserve feed. These "artificial" measures would raise the carrying capacity, as Smith defined it, toward its "natural" potential by enabling more animals to survive drought periods. The concept of carrying capacity thus elided the distinction between range and pasture, between natural and modified conditions; it confidently asserted that the natural variability in forage production could be smoothed out by technical know-how. Bentley's work appeared to support this optimism.

EARLY RANGE RESEARCH IN ARIZONA

In January 1901 David Griffiths, a botanist at the Arizona Agricultural Experiment Station in Tucson, undertook the first range restoration experiments in Arizona on a 240-acre parcel of land withdrawn for the purpose at the request of the Secretary of Agriculture. (It is today part of the Davis-Monthan Air Force Base.) Taking careful note of elevation, topography, soils, rainfall, temperature, and vegetation, Griffiths fenced the parcel and subdivided fifty-two acres of it into sixty plots, where he sowed forty species of forage plants under a variety of cultivation methods. He also planted a grass garden at the University of Arizona, using seeds obtained both locally and elsewhere. That summer he transferred to the new Bureau of Plant Industry, but he continued his research. Two years later, the USDA established the Santa Rita Experimental Range on Forest Reserve land southeast of Tucson. Almost 31,500 acres in size, the Santa Rita Range offered Griffiths a more representative tract of Arizona rangeland to study. Again, he fenced the area off from cattle and carefully documented its soils, topography, rainfall, and vegetation.

"The [native] perennial grasses have been completely destroyed on large portions of the range," Griffiths wrote in his first report (1901: 31). Nevertheless, he was optimistic that methods of cultivation could be developed to reestablish perennial grasses more quickly than they would do on their own. He was working during the 1898–1904 drought, however, and his results were not encouraging (1904: 61–62). Native perennials, which even under natural conditions do not germinate easily, could not be seeded with enough success to justify the cost; disking had actually decreased the carrying capacity in experiments in Arizona. "Where the carrying capacity of the lands is low no methods of eradication of weeds will pay for the labor involved. All that can be done is to get out of the land all that it produces of valuable plants without the abuse of overgrazing and to utilize the weeds if it can be done, if not by cattle then possibly by sheep or goats" (1907: 22). As a final irony, Griffiths conceded that the handful of non-native species that had spread abundantly had all been introduced by accident.

If the range could not be artificially reseeded, then the issue of its natural dynamics in relation to grazing was all the more important. How many animals could the range support, and how could a rancher maximize use without doing damage? Here Griffiths faced a conundrum. Range productivity varied from year to year, and recovery from past abuse was rather unpredictable. Annual plants were poorer than perennials as forage, but they reliably produced large amounts of seed, increasing the likelihood that some would germinate even under adverse conditions. Perennials produced seed only in good years, meaning that *consecutive* favorable years were necessary to enable them to spread significantly: one for seed production and one for germination. Since the establishment of the Santa Rita Experimental Range, only the years 1907–8 had met this criterion. Griffiths observed that "the natural restocking of the perennial range by new plants takes place at irregular intervals" (1910: 12).

The key variable, he concluded, was moisture. Rainfall was more important to range restoration than any humanly controllable variable or technique (1910: 14). Only wetter areas (at higher elevations or in drainages) showed promise for reseeding, and then only with annuals and nonnative species. Because of the variability of rainfall, however, these findings cast doubt on the very concept of carrying capacity. There simply was no static number of animals that a given piece of land could support in abstraction from time. Griffiths seems to have recognized this as early as 1904, when he wrote:

> Before any rational adjustment for the proper control of public grazing lands . . . can be made, much should be definitely known regarding the amount of stock that these lands will carry profitably year after year. This must form the basis of all equitable allotments. To secure such information is a most difficult task in a region where the seasons, the altitude, the slope, and the rainfall are so variable. It can be determined very easily in the Great Plains region, where conditions are uniform and reasonably constant, and indeed it is very definitely known there; but here the case is very different. (1904: 32)

By 1910, however, it was too late for Griffiths to call carrying capacity into question. To do so would have pulled the conceptual foundation out from under leasing, the cornerstone of range reform. Griffiths knew that his conclusions would be used to determine the economical size of allotments and that this required fixed assessments of carrying capacity. He thus set about to determine carrying capacity as though it were an atemporal attribute of the land. At twenty-eight locations on the Santa Rita Range, he measured the size and weight of all forage plants produced in the winter and summer growing seasons. After subjecting the data to a series of calculations, he determined that thirty-seven acres of range would support a cow for a year (or about seventeen head per section). Elsewhere in the region, carrying capacities were lower—fifty to one hundred acres per cow per year. He qualified these findings with several general statements urging conservative stocking. "Results seem to be secured much more rapidly by proper protection from overgrazing than any other method" (1910: 24); "the perennials . . . will again regain their ascendency over the weedy annuals when given a measure of protection" (1910: 16). Comparing his numbers to earlier data, he concluded—in conflict with his own analysis and eerily reminiscent of Bentley's findings— that three years' complete rest was sufficient to restore the range to its "original productivity" (1910: 24).

By today's standards, Griffiths's measurements of carrying capacity were rather high, but the pressures of World War I compelled the USFS to allow still more cattle onto the range. It was expedient in this context to find scientific support for higher stocking rates. In 1916 E. O. Wooton published further results of experiments comparing grazed and ungrazed areas of the Santa Rita Range. He found that perennial species continued their gradual displacement of annuals, especially at higher elevations, and that well-managed grazing slowed but did not reverse this trend. He calculated forage production at eleven hundred pounds per acre per year, almost all of it the result of summer season growth, and he suggested that no more than 60 percent of this growth could be sustainably grazed. The carrying capacity was therefore one cow per twenty acres, or thirty-two head per section, almost twice Griffiths's figure. In Wooton's report, Griffiths's insights into the wide temporal variation of desert grassland productivity are virtually absent.

The World War I era stock increases undermined the USFS's understanding of itself as dedicated first and foremost to natural resource conservation; it would be many years before carrying capacities were effectively enforced on the ground. Natural variability made this both more difficult and more important. In good years, lessees could point to ample forage to argue for higher stocking rates, but in bad years even the authorized numbers could do damage. Off the USFS land, in places such as the Altar Valley, these arguments did not even occur, although the challenges were the same.

THE CREATION OF STATE TRUST LANDS

After decades of anticipation, Arizona was authorized to become a state by act of Congress in 1910. A constitution was drafted and ratified by voters in the general election of 1912. Under section 28 of the Enabling

Act, the new state was deeded four sections of land per township in trust for the express purpose of generating funds to support public institutions (primarily schools). To this end, provisions were included to restrict disposal and management of the State Trust lands: They shall not be mortgaged or encumbered; they shall only be sold at appraised, fair-market value; any sale or lease shall be advertised and open to competitive bid, with the exception of leases of ten years or less.[4] The overriding principle of the restrictions and the intent behind them were to compel the state to put the lands to their "highest and best use" in behalf of the trust. As trust law has evolved over the course of the century, this has come to mean the same thing in public trusts as in private ones: that the trustee must strive to maximize revenue (Souder and Fairfax 1996). This appears simple enough and far simpler than judging "highest and best use" by aesthetic or environmental standards. No temporal period was stipulated, however, making the goal of maximization subject to broad interpretation. Should the State Land Department attempt to maximize revenue immediately by liquidating the lands and investing the proceeds, or should it hold on until a later time? This ambiguity gave the Land Department some discretion in managing its lands, but at the price of exposing it to political pressure from interested private parties, such as ranchers and, in a later period, developers.

A second historical circumstance complicated State Trust lands even further. In theory, Arizona was to receive sections 2, 16, 32, and 36 of every township in the state. By the time of statehood, however, a great deal of land had already been withdrawn from the public domain through homesteading, Indian reservations, national forests, grants to railroads, and mining claims or for reclamation under federal irrigation projects. A State Selection Board was therefore established under the Enabling Act to choose in-lieu sections from the remaining public domain to make up for these. It was through this board that the majority of the Altar Valley became State Trust land.

Having long capitalized on free public range, by 1910 cattlemen such as Jack Kinney recognized the desirability of securing legal, exclusive grazing privileges. The federal government would not close the range on its nonforest lands until the Taylor Grazing Act of 1934, so for more than twenty years the State Selection Board was the only mechanism to secure economically viable areas of rangeland (that is, larger than the half-sections permitted under federal homesteading laws). The result was a flurry of legal and bureaucratic activity, with cattlemen hiring brokers in Phoenix to shepherd the paperwork through the process of withdrawal, selection, and assignment of leasehold. Selections by the board had to be made one section at a time through the General Land Office (GLO). It appears that the GLO handled these requests in the pattern of homesteads, insisting that each section be linked to a separate individual's name. The result, for a time, was a huge number of small leases: The J. M. Ronstadt Company helped secure 103 state leases for La Osa in the year 1917 on less than one hundred sections of land. The status of sections yet unclaimed bewildered managers on the ground, and neighboring ranches disputed boundaries and fences well into the 1920s. (In one case, La Osa feuded with the same Ronstadt family that had earlier brokered its leases.) Leases were soon consolidated into larger, more easily administered blocks.

The constitutional mandate to generate revenues from State Trust lands conformed well to the demands of capital invested in ranching. Allowing cattlemen and their brokers to perform the central duty of the State Selection Board—selecting lands for inclusion in the trust—ensured that selected lands would be promptly leased, initiating a revenue stream. It also effectively ratified the informal "range rights" of the boom period: Who else but the nearest operator would choose to lease lands that generally lacked water? The exemption of short-term leases from advertisement and competitive bidding—inserted into the Enabling Act and later amended at the behest of the livestock industry—likewise reflected economic necessity. Inasmuch as grazing lands required improvements to be remunerative even for livestock, ranchers (and especially their bankers) insisted on long-term tenure guarantees as a condition of making these investments. (Under a 1951 amendment to the Enabling Act, private ranchers were guaranteed property rights in the improvements they made on trust lands.) This, in turn, allowed underwriters to view leased trust lands as security for loans, very much like deeded acres. For purposes of incorporation, for instance, Kinney valued La Osa's leases at sixty cents per acre.

One unintended consequence of the selection process was that state lands came to encompass a disproportionately large amount of Arizona's finest grasslands, especially in the southeastern and central parts of the state. Today, the mild climate and gentle terrain of these lands make them some of the most desirable and easily developed areas in the state. Unlike most federal lands, State Trust lands may be privatized. Since 1960 the State Trust has earned large increments from the increasing value of its lands for residential real estate (cf. Arizona State Land Department 1997: 32).

LA OSA LIVE STOCK AND LOAN COMPANY

In 1915 Jack Kinney incorporated the La Osa Live Stock and Loan Company, with capital stock of $5 million. At the time, La Osa comprised 20,000 acres of deeded land and 130,000 acres of state leases, bounded by three hundred miles of fence. Together with about 60,000 acres of unfenced public domain, these lands were grazed by eighty-two hundred cattle. Kinney became president of the new company; Ramon Elias and José Camou of Nogales became vice-president and secretary-treasurer, respectively. Elias and Camou were principals in the West Coast Cattle Company, which ran cattle in Sonora and Arizona and maintained offices in both countries. The new La Osa letterhead listed the five ranches consolidated by William Coberly—La Osa, Palo Alto, Pozo Nuevo, Secundino, and Buenos Aires—while proclaiming "1,000,000 acres under control," a figure that was either inflated or included lands controlled by the West Coast Cattle Company in Mexico.

The new part of the name—Live Stock and Loan Company—is deceptively simple compared to the business activities it signaled. The new La Osa combined features of ranching, banking, international trade, and futures trading. Kinney made loans to ranchers in the area, frequently secured by cattle; the growth of the animals on the range raced against the compounding interest on the loan. If the cattle grew (or produced as calves) more than, say, 8 percent per year *and prices did not decline,* the rancher made a profit. In good years, a herd could easily win this race, and strong prices made these loans reasonably low-risk during the war period. The "live stock" part of the business was basically identical, except that Kinney and his partners were the borrowers. Kinney's greatest asset was his access to the capital needed to make the whole operation possible. The Merchants National Bank of St. Paul, Minnesota, was his major backer, and the Consolidated National Bank of Tucson and the North Texas Trust of El Paso played lesser roles. Many of the private investors who purchased shares in the company were board members of the banks.

The productive basis of La Osa's activities was cattle: grazing, growing, and calving on the range. The profits (or losses), however, had more to do with moving cattle around, both in space and in time. La Osa shipped animals from Mexico into Arizona and thence to New Mexico, California, or the Great Plains. Camou and Elias oversaw the first step while Kinney traveled around the country buying and selling cattle, coordinating contracts and deliveries, and conferring with underwriters. To capitalize on time, Kinney could contract to buy or sell animals at an agreed price (per hundredweight) sometime in the future, hoping to profit on both the growth of the animals and a change in prices between contract date and delivery date. The La Osa lands in the Altar Valley served as a giant holding facility for this highly speculative enterprise. Camou wrote to Kinney in St. Paul in November 1915: "We are glad to know that you have orders for from five to six thousand head of ones and twos [one- and two-year-old animals] for spring delivery. No doubt we can buy them cheaply right now, and we believe that we could take care of about 3000 at our La Osa pastures with the present condition of the grass." The main office of the company was in Nogales near the port of entry; it shared a Post Office box with the West Coast Cattle Company. Subsequently, La Osa constructed a dipping vat near Sasabe so that cattle could be imported directly onto the ranch.

The critical elements for La Osa's bottom line were market prices and the growth of cattle. Kinney could, and did, squabble with buyers over prices and with railroads over freighting charges, but market prices were generally beyond his control. In the race to grow cattle at a rate higher than interest, rain could make the difference between winning and losing. Without rain, there could be no good feed; without feed, cattle could not gain weight or breed. Virtually every letter sent to or from Kinney closed with a brief discussion or inquiry about whether it had rained recently, how much, and where. It was often the only thing about the ranch itself that warranted comment, all else being legal or financial affairs.

Examined over a period of years, these remarks tell a story of wide fluctuation, with disaster and salvation sometimes only a few weeks apart:

3 NOVEMBER 1915: "We have not had any rain at all, and cattle is *[sic]* getting thinner."

17 AUGUST 1916: "Raining here most every day and our ranches are fine and we will have plenty of grass for three times the cattle we have in stock at present."

17 FEBRUARY 1917: "We have had no rains since early in January and conditions in some sections are not good."

28 JUNE 1919: "The range is in critical condition, and unless timely rains come along we will have the usual big loss."

18 JULY 1919: "It has rained every day since June 30th on some part of our range. . . . Experienced ranchmen tell me there never has been such an abundance of rain or such excellent prospects for fall and winter feed in fifteen years."

23 APRIL 1921: "It is very dry here. . . . Our cattle are very think *[sic]* but the death loss has been extremely light."

28 JUNE 1921: These are "the worst drouth conditions that have ever been known."

23 JULY 1923: "The rain so far has been local in character, but the feed started well over part of our property visited, with the exception of a strip about three or four miles wide, extending across the entire valley. . . . It has been raining somewhere around Tucson every day since we made the trip and we are in hopes it is reaching all parts of the property."

The obsession with rain was understandable, especially given the high stocking rates that La Osa maintained. Every pound of forage was potentially important to the bottom line. Actual stocking no doubt fluctuated widely, but it appears that nine thousand head was about the minimum and twelve thousand was closer to the average during La Osa's prime. In a 1921 letter to a potential buyer, Kinney claimed to have run as many as twenty thousand head at one time. These figures suggest that stocking rates ranged from about twenty-five to sixty head per section, with an average of roughly thirty-five, which is very close to Wooton's 1916 estimates of carrying capacity for the Santa Rita Experimental Range. In higher-elevation areas of the valley (where more rain usually fell), stocking rates were still higher. In 1917 Kinney's ranch manager reported that the neighboring Los Encinos Ranch, which they were negotiating to buy, was carrying eleven hundred head on ninety-two hundred acres (more than seventy-five head per section). "This range is in excellent condition and should carry 1,000 head of cattle," he judged.

In years of good rainfall, the valley's grasses appear to have supported these high stocking rates. But the perennial grasses that dominated before the droughts of the 1890s were not coming back; instead, annual grasses were taking their place. In 1917 Kirk Bryan surveyed the Altar Valley for the U.S. Geological Survey (USGS), focusing on water sources and geology. He later produced the most detailed accounts of vegetation that we have for this period (Bryan 1922, 1925). Of the La Osa area he wrote, "The predominant plants are grasses, mostly of annual types, which spring up after rains, seed quickly, and then cure while standing. . . . Occasionally on the plains and usually at the borders of the mountains are clumps and areas of perennial grasses" (Bryan 1925: 375). Prominent among the annuals was six-weeks grama *(Bouteloua barbata)*. Bryan hypothesized that higher rainfall promoted grasses in the portion of the valley above three thousand feet and that the grasses in turn supported fires, which served to control the spread of mesquite and other woody species. (He did not provide any direct evidence of fires, however.) "The mesquite," he wrote, "is present . . . but . . . restricted to the stream channels and to broken country. The bushes are, however, low and stunted" (Bryan 1925: 375). It seems that taller trees had not yet had time to reestablish after the fuelwood harvesting of the 1890s.

The distribution of grasses described by Bryan suggests that grazing played a causal role in the shift to annuals, mediated by water improvements. La Osa owned sixteen of the twenty-two wells in the valley. When the State of Arizona required that surface water rights be registered in 1919, Kinney filed on 305 separate rights to hold water in more than two hundred earthen tanks. When full, these tanks afforded water within two miles of almost any point in the valley. In striking contrast to pre-1880 conditions, Bryan observed that "[i]n the Altar Valley . . . stock-watering places are so numerous that the traveler will have no difficulty in obtaining water." He added, however, that "additional water supplies near the mountains are necessary" to utilize all the available forage (1925: 256, 254). It seems that perennials generally persisted only in areas where water had not been developed and thus where grazing had been limited or impossible.

The Altar Wash was well established by this time but not yet large enough to impact vegetation in the adjacent floodplain. Bryan described it as "a well-defined arroyo with banks from two to six feet in height" (1922: 73). He also observed "occasional hay fields" in the north end of the valley, and Kinney reported "several hundred acres of good land that sub-irrigates" in 1921. These were probably the same areas. In 1880 they had been dominated by giant sacaton, but by the late 1920s they had been sown with (or adventitiously invaded by) Johnson grass *(Sorghum halepense),* which has very similar habitat needs. Kinney and his associates did not see the wash as a problem but rather as an opportunity to create more reservoirs like Aguirre Lake. "We have already finished near Palo Alto, a big dirt dam and have under construction two more, near Secundino and Pozo Nuevo," Camou wrote to Kinney in the summer of 1916. "Our idea is to try and store plenty of flood water in every one of our ranches in order to save the expense of pumping water out of wells."

In July 1917 the problem was not too little water but too much. In Tucson 3.9 inches of rain fell (Bryan 1925: 40); the rain gauge at La Osa headquarters recorded more than eight inches. "I never saw a rainy season like this before," the ranch manager wrote to Kinney. "The telephone line is badly washed out, most of the valley fences are down, both bridges at Palo Alto are washed away, and an automobile bogs down on any side hill." (The presence of bridges at Palo Alto indicates that entrenchment had reached the north end of La Osa's lands by this time.) Most of his letter recounted events at Aguirre Lake:

> If you want to see some real excitement you should see the fight I have had for the dam. First the concrete flood gate was taken out by the roots, but we were there with teams and men, and managed to get enough dirt in behind it to hold the dam. Yesterday, the water cov-

ered the corrals (not the tops) at Buenos Aires and the shooting cabin was out of sight. The water reached within 18 in. of the top of the embankment and driven by high wind the waves were actually washing over in places. The top of the dam was badly washed and only 8 in. thick in places. I had all the men and teams we could raise working on the weak places, but figuring it couldn't last another hour at that rate, I took a gamblers chance and cut away a section. You know what it means to have a dirt dam start with five billion gal[lon]s behind it. I had lots of brush and stakes ready, and even put all the bystanders who came down to see the dam go out, to work. I cut at 11 a.m. and at 6 p.m. we had her harnessed and under control again.

What effect this barely controlled release had on the Altar Wash can only be imagined. Photographs from 1928 document a much larger and deeper channel (see chapter 4), and this quasi-natural flood event may well have caused much of the growth. Presumably the newer dams downstream collapsed. Bryan (1925: 147ff.) does not describe the flood, but he credits the La Osa manager with a special design for dams and spillways using interlocked logs and mesquite branches to fortify the earth. "This device, installed by Mr. Kibbey [the ranch manager] in the dam of the lake at Buenos Aires, has been in successful use for a number of years" (1925: 150).

As the example of the dams suggests, the purpose of improvements was not so much to fix the damage of past years as to mitigate the uncertainty surrounding rainfall and forage growth. Having lots of water tanks increased the chance that at least some of them would fill, in spite of the area's spotty and unpredictable summer rains. This, in turn, helped ensure that grass would be available near water in some portion of the range. Cutting and stacking Johnson grass in the old floodplain, as was done on the Palo Alto Ranch through the late 1920s, provided feed for times when the range was inadequate; in very lean times, cottonseed cake or other supplemental feeds were delivered to the ranch to keep the cattle growing. Consistent with Smith and Bentley's recommendations, artificial measures aimed to stabilize conditions during poor years and thereby make "carrying capacity" meaningful over time. Ideally, natural processes would suffice (filling tanks with runoff, for instance) so that more expensive measures (burning fuel to pump water from deep wells) could be avoided. Substituting human labor and inputs for natural bounty was expensive.

Early on, Kinney and Camou sought to economize by squeezing more out of their employees. The ranch employed some three dozen men (mostly cowboys, farmhands, and pump men), each making fifty cents to about two dollars a day in 1917. "We are endeavoring to establish an efficient force in all our departments and we expect to succeed even if it is necessary to fire half of our employees in order to obtain it," Camou wrote to Kinney in 1916. The Mexican Revolution contributed to an atmosphere of general turmoil along the border; cattle frequently disappeared across the border or were slaughtered on the sly for food, and employees were sometimes complicit. Kinney complained repeatedly about lazy, untrustworthy, or reckless foremen and hands, both Mexican and Anglo. To reduce labor costs La Osa established a company store, where most employees spent their wages even before receiving them. The store also profited from local residents (mostly Mexican smallholders and Indians from the reservation), who bartered livestock for staples such as flour and beans.

Compared to market forces and the weather, however, human labor was a minor factor in company prospects. The pressures of debt were continuous and steady, as were the needs of the cattle, whereas nature produced its bounty on an irregular schedule. Dry conditions and the postwar agricultural contraction crippled La Osa after 1918. That year the company's books showed some $420,000 of debts on assets of just less than $700,000. A receiver was appointed in 1919 after a major shareholder filed suit, and in February 1920 the company was reorganized, drawing still another loan from Merchants National Bank to buy out the interest of the North Texas Trust. Losses for 1919 totaled $137,000, and liabilities were more than three-quarters of assets. Severe drought the following summer only made matters worse, on the books and on the ground. Cattle prices dropped precipitously, making ranchers reluctant to sell. "In 1920, livestock numbers were high and the animals were restricted by the new fencing. The summer was extremely dry and there was no market for livestock. Range areas formerly not used heavily were so for the first time. This year probably saw the start of the large scale change of the valley's flora from grasslands to shrublands" (Robinett n.d.: 3). On the neighboring Santa Margarita Ranch, dead cows littered the ground around stock tanks.[5] Another reorganization of La Osa took place in 1923. By 1926 assets had fallen to $338,000 against liabilities of $320,000, and Kinney was actively seeking buyers. He sold the Secundino, Buenos Aires, and La Osa Ranches to Fred Gill and Sons in October 1926 for $75,000. Two thousand cattle accompanied the seventy thousand acres of land, suggesting that the stocking rate had dropped by about half, to eighteen head per section.

Jack Kinney held on to the Palo Alto Ranch until 1931, when he moved north to a ranch near Red Rock. Campaigning for the governor's office in 1932, he played up his image as a colorful cattleman. He disbanded the Live Stock and Loan Company, but the La Osa name persists. In late 1927 the Gills sold the La Osa Ranch headquarters and eight hundred acres of land to Arthur Hardgrave, who owned a string of ice companies in Kansas City. With his wife, Hardgrave converted La Osa into a dude ranch, catering to eastern socialites in search of an authentic western experience. They bought the dipping vats for color, Mrs. Hardgrave told reporters. Guests to the ranch assembled on the lawns to watch as large herds of cattle were brought across the border, dipped in chemicals, and herded away.

CONCLUSION

The La Osa period was defined by the struggle to reconcile two temporalities, one ecological and the other economic. The first was unpredictable and highly variable, defined by irregular pulses of rainfall and drought. The growth and reproduction of forage grasses and livestock pulsed to this rhythm (Pieper et al. 1983). Research conducted elsewhere in the region suggests that the range may have produced ten times as much forage in "good" years as in "bad" ones (Herbel and Gibbens 1996). The second temporality was abstract and regular, defined simply by the clocklike passage of time. Interest grew and compounded; dividends and debts came due at set dates on the calendar. Kinney and his associates endeavored to convert the pulses of natural growth into profits in the marketplace, but the greater the burden of their debt, the more difficult (and imperative) this became. Overstocking during drought was an almost inevitable result. Herds built up during rainy periods could not be supported on the range, but low prices, which always accompanied drought, meant that selling the cattle was tantamount to defaulting.

Early government researchers sought to aid in this reconciliation by fixing the spatial coordinates of range livestock production in terms that were independent of time. All observers agreed that the open range had been disastrous, and dividing the range into exclusive leased allotments seemed the obvious solution. In theory, carrying capacities and lease systems worked together, translating natural phenomena into abstract terms amenable to both bureaucratic regulation and market exchange. Lease systems could only realize this ideal, however, by assuming that natural conditions were reducible to an atemporal calculus: allotment x would support y number of cattle every year. Just as the Homestead Act enshrined the norm of 160 acres for a farming family, the formation of ranching envisioned that any given amount and type of land ought to support a specified number of animals, and for the same reason: to allow "the use of enough land to support a home" for "a single family" (Potter 1905: 26). Bentley and Griffiths's claims that lost productivity could be recovered in three years likewise abstracted from the vagaries of particular, actual years.

The abstraction of range science from historical time was institutionalized between the wars, following publication of Arthur Sampson's *Plant Succession in Relation to Range Management* in 1919. Earlier researchers such as Smith, Bentley, Griffiths, and Wooton had assumed that the range would return to its earlier conditions if rested from grazing, but they had lacked a theoretical model to explain how or why this might occur. Sampson, a USFS plant ecologist, filled this need by adapting Frederick Clements's theory of succession to rangelands managed for grazing. Sampson's model rested on two assumptions, one borrowed directly from Clements and one derived from his own observations in the Wasatch Mountains of Utah. The first held that the "stages" through which plant communities had evolved over geologic time—from bare rock through lichens, weeds, perennial grasses, and trees—constituted a more or less fixed sequence of successional stages determined by soil and climate; the "natural" tendency was linear and progressive toward a single, stable climax community.[6] Sampson's contribution was to argue that the successional sequence would recur on shorter temporal scales following disturbances such as overgrazing. Removal of grasses—and in cases of severe erosion, topsoil as well—was analogous to moving back in time, "resetting" the ecological system to an earlier stage. (The Buenos Aires Ranch during the La Osa period fit this model neatly: Grazing had pushed the plant community back one stage, from perennial to annual grasses.) By implication, grazing pressure had an effect exactly opposite to natural succession. It followed that at some theoretical stocking rate, successional and countersuccessional forces would balance, holding a vegetation community at a desired stage for maximum productivity. At midcentury, this model achieved quantitative formulation and solidified into the theoretical core of applied range science (Dyksterhuis 1949).

The resulting management paradigm presumed that the products of ecological processes—grasses and livestock—could interface with the market just as the products of industrial processes did, with production measured against time but no longer dependent on it. The range could be managed for desired conditions by careful selection of the proper stocking rate. Grazing was typically continuous year-round; in southern Arizona there was no need to move between summer and winter grazing. Each pasture was expected to support a more or less fixed number of animals; the herd and the carrying capacity it represented were both capitalized assets. It was as though the range were a beef factory. Based on this model, ranching aspired to eliminate pastoralist dependence on nature, even as it sought whenever possible to let nature do the work. In later chapters, I will examine other dimensions of this joint public-private endeavor, such as cattle breeding and range restoration.

In practice, however, the formation of ranching remained incomplete. Stock tanks, wells, and fences could improve the distribution of grazing and minimize losses during droughts, but they could not liberate range cattle production from its ultimate dependence on the natural processes of rainfall and forage growth. Even when the limitations of carrying capacity were acknowledged, the concept remained central to research, policy, and management. In a 1905 report to the Senate, for example, Gifford Pinchot's newly minted Forest Inspector Alfred Potter wrote, "The ranges have as a rule been stocked on the basis of their capacity during normal or good years, and consequently whenever there came a season of drought they have been found to be overstocked" (Potter 1905: 17). Yet this same report prepared the way for leases on USFS lands, using fencing, improvements, and measurements of carrying capacity.

In the realm of popular culture, meanwhile, the abstraction of ranching from time and history was a stunning success. Until the 1890s cowboys were widely considered dissolute, whoring renegades, and cattlemen imperious land barons. By the 1920s the two figures had collapsed into a singular, mythologized image of rugged individualism, frontier chivalry, and nature-conquering courage. It was this mythology of ranching that drew guests to dude ranches such as La Osa; that Kinney (and many others) cultivated in political campaigns; and that found further expression in countless western novels, movies, and tourist attractions.

Like myths everywhere, the mythology of ranching resides outside of time and history. It romanticizes pastoralist elements of the boom period: cowboys trailing herds across fenceless ranges; confronting rustlers, "nesters," and Indians; and leading lives of freedom and adventure in the wide open spaces of the West. The fact that such practices endured for only a generation has not prevented them from becoming timeless American legends. Even the role of state agencies in the formation of ranching has been selectively mythologized in favor of its pastoralist dimensions: the Arizona Rangers policing the range in pursuit of rustlers; the U.S. Army battling Apaches to secure American possession of land and livestock; and the advent of standardized livestock brands, in combination with laws making it illegal to sell cattle without the signature of an official brand inspector. The rugged, violent, antigovernment mythology of ranching dates to a time when citizens had no choice but to enforce private property relations themselves because the state could not. Other state activities—leases, carrying capacities, and scientific research—have remained relatively obscure despite their greater effect on postboom range ecology and management. Unrelated to pastoralism, they seem less suited to mythology and legend; they reveal private production and public policy not as opponents but wrapped in a mutual embrace.

What we mythologized, in short, was precisely *not* ranching, though we call it by that name. It was decoupled from actual, on-the-ground practices from the very start. This misrecognition enabled ranching to acquire its particularly American symbolic value, first among the eastern and British financial classes and subsequently in popular culture. For investors, the idea of wealth as a product of nature appeared to resolve "the central contradiction of the finance form of capitalism": how to realize value without producing it (Harvey 1985b: 88). In British periodicals, "the running of a cattle ranch was pictured as something of a sporting affair, for the 'ordinary work consists of riding through plains, parks, and valleys.' There were, it was said, only two short seasons for hard work 'when masters and men, well mounted,' carried on a roundup" (Graham 1960: 423–424). Profits from cattle were constructed as "natural" whereas losses were not.[7] For the middle and working classes, too, growing rich without working was a tantalizing fantasy, as was the idea of working outside of the boss's presence. The two rarely coincided in practice, but from a distance this fact was inconsequential.

What the scientists, bankers, and dudes had in common was their vantage point: They apprehended "ranching" from the perspective of industrial capitalism. Consider again the issue of time: "Two apparently contradictory themes permeate the entire literature on pastoral societies. One tells of the arduousness of the herdsman's existence, conveying an impression of unremitting toil and frequent physical hardship. The other remarks on the leisurely pace of life of the pastoralist, who has only to look on as his animals seek out their food and multiply of their own accord" (Ingold 1980: 180). This apparent contradiction characterized the mythology of ranching from very early on. Teddy Roosevelt, in his popular *Ranch Life and the Hunting-Trail,* wrote this about roundup: "[E]very one feels that he is off for a holiday; for after the monotony of a long winter, the cowboys look forward eagerly to the round-up, where the work is hard, it is true, but exciting and varied, and treated a good deal as frolic. There is no eight-hour law in cowboy land: during round-up time we often count ourselves lucky if we get off with much less than sixteen hours" (1888: 54–55). Similarly Joseph Munk, in his *Arizona Sketches,* wrote, "Every day has its duty and every season its particular work, yet there are times of considerable leisure during the year" (1905: 62). There is no contradiction in these claims if viewed from the vantage point of pastoralism. The work involved in keeping a herd of free-roaming cattle domesticated, for example, consists largely in continuous physical copresence, even if the herdsman is doing little more than observing his animals. It is not a question of more or less work, but rather of the meaning (or meaninglessness) of the concept of "work" itself (Ingold 1980: 182).

The work of ranching promised an escape, both from urban industrial settings and from the social relations they entailed. Ranching could appear outside of time precisely because it remained fundamentally pastoralist, irreducible to the abstract clock time of industrial capitalism (Postone 1993; Thompson 1967). As Roosevelt (1888: 6) put it, "Ranching is an occupation like those of vigorous, primitive pastoralist peoples, having little in common with the humdrum, workaday business world of the nineteenth century." A life of unmediated immersion in nature struck Roosevelt, Munk, and countless others as more authentic, honest, and vigorous than industrial life. Features of the landscape—fresh air, expansive vistas, wildlife, and sunsets—acquired a symbolic value tied to the consciousness and bodily condition of the (male) individual. On the ranch one could be outside of "civilization," *be aware of that condition,* and enjoy it by virtue of the contrast. George Henderson (1999: 105) has argued that irrigated agriculture in California during the same period offered wage workers "the opportunity to become the capitalist who still remains a laborer." Ranching offered capitalists the opportunity to become (or pretend to become) the laborer while remaining a capitalist. It combined and affirmed, however paradoxically, both the "rugged individualism" of the frontier and the financiers' fantastic visions of wealth without labor and profits without class struggle.

The mythology of ranching picked up where and when the agrarian myth of the yeoman farmer had foundered: on the arid and semiarid plains in the last decades of the nineteenth century (Smith 1978). It replaced agrarian symbols (the plow, the soil, and the toiling farmer) with pastoralist ones, only to misrecognize them as ranching. Much like the mythical Jeffersonian farmer, the composite cowboy-rancher figure could appear free, democratic, and prosperous, regardless of actual conditions, only by collapsing the distinction between capital and labor. In this case, however, the distinction did not have much purchase to begin with: The decisive class relation was not between rancher and cowboy—as significant as that relation was to the individuals involved—but between lender and borrower, investor and rancher.

A more rigorous definition of ranching reveals both the historical specificity of the cattle boom and the ideological potency of the myths that it spawned. The very factor that distinguished boom-era livestock production from present-day ranching—common access to land—provoked the influx of capital that made it so ecologically devastating. Subsequent ranchers struggled to mitigate that damage, simultaneously enabled and constrained by their dependence on credit. Notwithstanding its many cultural continuities, ranching proper is structurally distinct from the pastoralist-ranching hybrid of the boom. Unfortunately, scholars and mythmakers have collapsed the distinction between pastoralism and ranching by reducing both to a set of symbols: cowboys, spurs, big hats, etc. This culturalist confounding has obstructed cogent analysis of ranching in the American West. Its myths have thus persisted, even while twentieth-century ranching has endeavored to eliminate its pastoralist alter ego. The mythology of ranching has long since taken on a life of its own, both among well-to-do ranch owners and among the developers and home buyers who make up the flourishing market for suburban "ranchette" properties.

producing Nature

Fred Gill and Sons owned the Buenos Aires Ranch for thirty-three years, longer than any other owner. Their tenure spanned the Great Depression, the Dust Bowl, World War II, and the severe drought of 1951–56. Each crisis helped to ratchet up the role of state agencies in ranching within the Altar Valley as elsewhere in the West. Although the business fortunes of the Gills are known in less detail than those of Kinney, this lack is compensated by a greater array of government reports and personal recollections than for the earlier period.

The continued dependence of ranching on highly variable natural processes did not disrupt the paradigm of research and management established between the cattle boom and World War I. To the contrary, economic and ecological setbacks only augmented the determination of ranchers and agencies to make range livestock production more tractable. Even more than Kinney, the Gills invested in improvements intended to "fix" the spatial and temporal coordinates of production on the Buenos Aires Ranch. Agencies, meanwhile, began to act on the fact—recognized early on—that research alone was insufficient: Many ranchers were not keeping up with the latest scientific knowledge, or at least they were not applying it on the ground. As the La Osa period illustrates, stocking rates were routinely in excess of recommended carrying capacities (however questionable the concept may have been in itself), and the State Land Department would not even assign carrying capacities for its lands until the 1950s. Crises triggered by overproduction and drought were seen not as indictments of the paradigm itself, but as evidence of the need to find "the best methods of getting the acquired information to the people" (Galloway 1913: 5).

New state agencies were created to bridge the gap between science and practice and to introduce "modern" methods to agricultural production. The Smith-Lever Act of 1914 authorized the creation of cooperative agricultural extension programs, jointly funded by federal, state, and county governments (hence "cooperative") and closely integrated with land-grant university colleges of agriculture (Rasmussen 1989). Pima County established an agricultural extension office in 1920. Its first agent was active in the Altar Valley around the time of the Gills' arrival. The federal Soil Erosion Service—formed in response to the Great Depression and the Dust Bowl and later renamed the Soil Conservation Service (and more recently the Natural Resources Conservation Service)—used its much greater resources to expand on the nascent county programs. Lacking the authority to enforce carrying capacities, however, these agencies tended to focus on mitigating damage through further range improvements such as fencing, irrigation, revegetation, and water development.

The framework of government research, extension, and "improved" production examined in this chapter became the prototype for agricultural development projects on Native American reservations and later in a wide array of "Third World" countries (Rasmussen 1989). When imposed on Native Americans, state formation of ranching was often repressive and generally ineffectual, as White (1983) and Lewis (1994) have shown for the Navajos and Tohono O'odham (Papagos), respectively. Projects aimed at pastoralist societies in developing nations have had similar results, prompting many scholars to question the ethics and efficacy of "development" as a whole (see, e.g., Agrawal 1998; Ferguson 1994). Yet very little scholarship has examined the genesis of this model in government activities aimed at U.S. farmers and ranchers, where methods were less coercive and outcomes more ambiguous.

In the Altar Valley, improvements realized many of their short-term objectives, only to be undermined by longer-term unintended consequences, many triggered again by drought. Improvements during the Gill era fell into two categories: those related to the land and those concerning cattle. The Gills invested heavily in fencing, water sources, and erosion control, with most costs shared by the SCS after 1937. The range rebounded from the droughts of the early 1920s, and it appears to have reverted toward "original" conditions in the 1930s. By the drought of the 1950s, however, mesquite trees had begun to spread into the upland mesas previously dominated by grasses. The exact causes and effects of this shift are still debated today, but it is clear that grazing, drought, and water improvements all played important roles. Mesquite encroachment, in turn, provoked further research by scientists from the USDA and the University of Arizona, whose findings would be used on the Buenos Aires Ranch in the 1970s. Improvements in cattle, meanwhile, rationalized the two principal biological processes on which production relied: feeding and breeding. Here the unintended consequences were more economic and sociological than environmental, at least when viewed from the perspective of the Buenos Aires Ranch.

COOPERATIVE EXTENSION

The effort to increase agricultural output during World War I drew early extension agents into virtually every facet of rural life and work, from crops and livestock to nutrition and "home economics." Although raising production was the overarching goal, extension programs aspired to improve rural society as a whole, implicitly (and, starting in the depression, explicitly) casting it as impoverished and socially backward. Progressive ideology held that prosperity would flow from the application of scientific knowledge, disseminated to the populace through persuasive, practical demonstration.

In the western states, extension directors recognized range livestock production as the economic foundation of an area encompassing 650 million acres. They proposed programs to promote improvements such as those that the Gills made on the Buenos Aires Ranch: fencing, water development, improved breeding, and supplementary feeding (Lloyd 1924). Whether the Gills took their cues from extension is unknown, but the parallels are unmistakable. The practices and discourses surrounding these improvements constituted an officially sanctioned and publicly subsidized vision of what ranching should be. The agricultural agent for Pima County was C. B. Brown. Three of his annual reports from the period 1928 to 1931 have survived, offering a detailed picture of his activities in the Altar Valley and elsewhere in the county.

By his own account, Brown initiated his work with an interest in irrigation "for the production of range supplementary feeds." He told the *Arizona Cattleman and Farmer* (9[53]: 2) in 1926 that "it is no longer possible to build up a good breeding herd and produce animals that will take well on the market, without raising supplemental food." Irrigation promised to resolve the tension between recurrent drought and the economic imperative to invest in improved breeds (see below). "A man may have an excellent herd of Herefords, and along comes a drouth" which forces him to sell. He "is frequently forced to rebuild his herd at home from a non-descript class of stock." With irrigation, surplus feed could be produced and stored "for years, if necessary," through stacking or new techniques of ensilage.

Floodplains, such as the loamy bottomlands of the Altar Valley, appeared perfectly suited to this vision, and for a time in the 1920s formation of an irrigation district was contemplated. "[A]lfalfa would grow through the year" with only minimal irrigation in the winter, wrote the authors of a preliminary report on the matter (Goldman and Burns 1923: 1). The spring dry season posed a challenge, they conceded, but it was actually shorter than the dry season in the San Joaquin Valley of California, where irrigated agriculture was booming. It was even possible that the Altar Valley could grow fruits and vegetables for the "high-priced market in Tucson" (ibid.: 3). Shortly before selling the Buenos Aires Ranch to the Gills, Jack Kinney cooperated with Brown on a survey of the bottomlands around the Palo Alto Ranch, with the goal of engineering dikes and ditches that would capture floodwaters for irrigation. Two years later, in 1928, Brown included photographs of large stacks of Johnson grass hay on the Palo Alto Ranch in his annual report.

The vision of floodplain irrigation in the Altar Valley was a watershed-scale version of what Pedro Aguirre had done behind his dam beginning in the 1880s. It remained dependent on good rains, but it was also vulnerable to too much rain, which threatened dams and could carry away the very bottomland soils that Brown sought to turn into fields. Over time, as he witnessed the growth of the Altar Wash, Brown came to see water as "one of our most important annual crops" and erosion as a threat to the county as a whole. "The growth around Tucson and the surrounding country since 1920 has made necessary the use of more water each year, and evidence already secured indicates that we are building up close to our water line" (Brown 1931: 3). Yet as much as 80 percent of water falling in major rain events was being lost. "[E]rosion has been so rapid during the past twenty years that a large percent of our floodwater is being carried from the confines of the county in channels much narrower and deeper than used

Figure 10. Contour plowing to reduce erosion in the north end of the Altar Valley, 1930. Measures such as this were promoted and funded by the government in response to acute erosion throughout the region. (From Brown 1931, p. 22a; courtesy of the Arizona Historical Society)

to be the case" (ibid.: 4). The result was low infiltration rates and declining water tables. This was of concern not only for ranches and farms but also for urban growth in general: "[N]ot least in importance in the financial growth of our county, has been the increase in value of foothill and mesa lands surrounding Tucson for residential purposes. . . . While such development does not require a large amount of water for its maintenance and continuance, an adequate and dependable supply is necessary to secure the confidence and investment of capital" (ibid.: 3).

Brown worked to draw public attention to the poor condition of watersheds in Pima County. He compared infiltration depths in areas with and without vegetation cover, and he experimented with contour plowing to increase water retention in the north end of the Altar Valley (fig. 10). He sponsored a contest in Tucson to raise awareness of the sediment load of floodwaters in the Santa Cruz River. He documented the Altar Wash and other sites of acute erosion in photographs. Despite initial difficulties enticing ranchers to demonstrations, he collaborated with several in revegetation experiments, dam construction projects, and the development of irrigated pastures for supplementary livestock feed. Numerous trials with non-native plants failed (including one on the Palo Alto Ranch using shad scale and Australian salt bush), but Johnson grass successfully colonized where (and when) moisture was good.

There was no question in Brown's mind that overgrazing was a major cause of watershed degradation in Pima County. "[M]uch of our watershed has been so heavily overgrazed and the surface loam soil so thoroughly removed that a large percent of the natural ground cover, especially grasses, has ceased to grow. In many areas even the more resistant types such as burro weed, cactus and mesquite, are dead or in a dying condition owing to the impervious exposed subsoil" (Brown 1931: 4–6). *"Grass and good ground cover [are] one of the cheapest and most effective water conservation measures on the range,"* he stressed. "We are, therefore, urging that, where practicable, the range be so fenced that controlled grazing can be practiced so that the ground cover, especially the grasses, will be left in a healthy condition" (ibid.: 6, emphasis in original). This equation of fencing with range conservation was, it appears, taken more or less for granted. Yet in a curious passage from three years earlier (1928: 24), Brown suggested the importance of other factors:

> There is a tendency on the part of some ranch owners to graze their range much too heavily. Some have largely ceased to maintain an extensive breeding herd. Their ranches are largely unstocked during the summer months. The forage crop thus produced is grazed off by steers during the winter months. How long this practice can be continued, I am unable to say but for the past two or three years it has been exceedingly profitable, and probably good for the range.

How could grazing "much too heavily" also be "good for the range"? Brown nowhere elaborates, but presumably it had something to do with timing: winter grazing followed by summer rest. Unfortunately, we do not know which ranchers practiced this management strategy or when.

Viewed historically, C. B. Brown's focus on erosion captures a paradox of lasting significance. On the one hand, he was among the first people to recognize the far-reaching implications of arroyos for watershed functioning and long-term sustainability. His work to mitigate erosion addressed a fundamental component of range degradation. On the other hand, in the alluvial zone of the Altar Valley the building of dams *presupposed* fairly high grazing pressure: Thick, ungrazed grass diminished runoff rates so much that reservoirs would not fill reliably. He praised Aguirre Lake, writing that "[s]uch bodies of water as this, aside from their economic value, add beauty and enchantment to the landscape" (*Arizona Cattleman and Farmer* 12[37]: 1). Yet the lake bore a large responsibility for creating the arroyo that Brown rightly impugned, and none of its values—economic or aesthetic—would persist in the absence of continued grazing.

THE GILLS

When Fred Gill bought the Buenos Aires Ranch, he intended "to maintain as high as six or seven thousand head of cattle on his newly acquired property," according to a local report (*Arizona Daily Star,* 12 October 1926). Gill owned a large and expanding cattle operation in the San Joaquin Valley of California. He also had three sons, and he resolved the problem of joint ownership by buying a ranch for each of them. The Buenos Aires became the domain of the eldest son, Roy. Until 1948 the Gills ran the ranch as a stocker operation, much like Kinney had done before. They purchased yearlings from the large Manning and Baca Float Ranches in the Santa Cruz Valley, grew them out on the Buenos Aires, then shipped them to ranches and feedlots they owned elsewhere in Arizona and in California, New Mexico, and Oregon. By the 1940s Fred Gill and Sons was one of the largest owners of cattle in the United States, supplying some sixty thousand head per year to the armed forces during World War II.[1]

As before, drought and depression alternated with periods of good rainfall and rising prices through the 1930s (Wagoner 1952: 58ff.). With their vertically integrated operation, the Gills were better positioned than most ranchers to weather these storms. They could move cattle to other ranches during drought or into feedlots, where concentrated inputs could be fed to thousands of confined animals prior to slaughter. This effectively displaced grazing pressure away from the Buenos Aires range (just as C. B. Brown envisioned) and onto the rapidly increasing output of irrigated agriculture in California and Arizona. The large herds on their ranches and especially in their feedlots provided a buffer, though by no means a foolproof one, against the ups and downs of market prices for cattle. Smaller operations had no such margins. In the 1930s the Gills bought out the Lopez and Garcia Ranches, smallholdings on the southeast border of the Buenos Aires Ranch. They also acquired pieces of the San Luis Ranch from the Boice family, the owners of the Arivaca Ranch to the east and several other large ranches in the region. With the addition of a USFS allotment in the San Luis Mountains, these acquisitions expanded the Buenos Aires Ranch to about one hundred thousand acres.[2] The Gills stocked the ranch at rates comparable to Kinney's: between twenty-five hundred and sixty-eight hundred head at any one time (fifteen to forty acres per head).

The Gills' legacy at the Buenos Aires Ranch was fences and waters. Employing a contract crew eight months a year, they installed 250 miles of new interior fencing at a reported cost of $250 per mile. In 1946 there were ten pastures; by 1959 there were sixteen. The posts were of Texas cedar milled on the ranch. Fences cut down on cowboy work, enabled seasonal rotation among pastures, and divided the ranch into four parallel grazing operations, each with its own manager, headquarters, and herd. Fences also gave ranchers a financial incentive to suppress range fires because the posts would burn.

In the late 1930s, tractors and bulldozers replaced draft animals and *fresnos* for earth-moving work, and the SCS actively promoted range improvements for erosion control. Through the SCS, the federal government shared the cost of building earthen stock tanks, fences, and spreader dams, among other things.[3] In 1941 the Gills mounted a sweeping program of water improvements. In most cases they modernized existing surface water tanks to maximize water retention and minimize the risk of blowouts. Rather than locate a tank on the main stem of a tributary, they displaced it to one side (or rerouted the main drainage), digging a trench above it to the wash. In the wash, they constructed concrete headworks with pieces of railroad track sticking up at intervals. When the tank was dry, they slid heavy boards between the iron tracks, closing the headworks and diverting water down the trench to the tank. When it was full, they removed the boards, allowing subsequent floodwaters to run down the wash. Often a single heavy monsoon would suffice to fill the tank with enough water to last a year or two (cf. Duncklee 1994). Multiple tanks were constructed along larger washes. Sand traps—depressions where the water would slow down and deposit its sediment before flowing into the main tank—were also installed at this time to prolong the utility of the tanks. By 1959 sixty-four reliable water sources dotted the ranch, most of them earthen tanks.[4] A crew of three contract employees worked year-round to clean, improve, and maintain waters. The Gills also installed a fourteen-mile, gravity-fed waterline from Arivaca Creek to Secundino, with water troughs at one-mile intervals. With water thus distributed and fences in place, only four cowboys were needed to oversee the entire ranch.

The Gills improved Aguirre Lake in similar fashion and on a much grander scale. Bailey Wash, which drains the southeast slope of the Baboquivari Mountains, was made to augment the lake by means of headworks and a long dike. In addition, a complex of dikes and spillways was constructed above the lake, creating a separate, deeper reservoir that could be filled from any of the three tributary washes. Finally, a pump and lock system was installed to regulate the whole works. The result was that water could be captured from any of three directions for impounding in the reservoir or spreading over an eighty-five-acre field of alfalfa. Water could also be released into the bottomlands below the lake to irrigate the Johnson grass and sacaton that grew there. In the event of a flood, the headworks on Bailey Wash could be opened; a spillway served to relieve pressure on the Aguirre Lake dam. The system extended more than three miles from end to end, though the elevation drop that allowed it to function was only forty feet. Combined with consistently heavy grazing in the surrounding watershed, the stock tanks provided reliable water throughout the ranch, while Aguirre Lake irrigated supplemental hay and bottomland vegetation even during drier periods.

The final component of the Gills' water program was spreader dam construction. These dams were meant not to impound water but to raise it up out of the entrenched arroyos and spread it across the old floodplain; the dams combined erosion control with ditchless irrigation. Earth was pushed up to form a dam set at an angle to the arroyo and extending out onto the adjacent flats. Floodwater would gather behind the dam, dropping its sediment in the incised arroyo. Through careful engineering, the overflow could be forced to move slowly out onto the adjacent plain; once it cleared the wing of the dam, it spread across the flats as sheet floods had done prior to arroyo formation. Kinney and the King family may have built spreader dams north of the Buenos Aires Ranch as early as the 1920s; the first of these dams in the southern Altar Valley were built on the neighboring Santa Margarita Ranch in the late 1930s. The Gills adopted the idea a few years later and, as was their habit, applied it on a far grander scale on the Buenos Aires Ranch. They constructed twelve to fifteen spreader dams on the Arivaca Wash between Figueroa and Secundino (a distance of four miles, in which the elevation drops a little more than one hundred feet). Thirteen of these structures appear on the 1979 USGS topographical map. They endured long enough to heal six to eight feet of downcutting; today the channel is only a couple of feet below the flats. (Farther upstream, where dams were never constructed, the bank remains fifteen feet high. Large cottonwoods in the arroyo bottom suggest that it is unchanged from the time of Bryan's observations.) Other spreader dams on Compartidero and Puertocito Washes were less successful. Presumably they worked as intended until flows became too heavy and washed the dams out.

Until the severe drought years of the 1950s, the composite picture of the Buenos Aires Ranch under the Gills' management is of an altered but economically productive regime of water, soils, vegetation, and livestock. The best evidence is rather general in nature, but it suggests that range conditions were good for cattle production through the 1930s and 1940s. SCS maps from 1938 depict the Altar Valley as divided along a northwest-southeast line passing through the north end of the Buenos Aires Ranch. Above this line, in the southern end of the valley, degradation of the range was "none to slight," whereas below it to the north damage was "severe." The *Ultra-extensive Report on Santa Cruz Watershed* (SCS 1936), completed during the same period, used a similar dividing line in reviewing range conditions and land uses. It estimated stocking rates of thirty to thirty-five head per section on "the best land" in the higher, southern areas of the Altar and Santa Cruz Valleys (p. 10). Vegetation in these areas was approximately 80 percent perennial grasses and 20 percent browse;

prominent grass species included slender grama *(Bouteloua filiformis),* curly mesquite *(Hilaria belangeri),* side oats grama *(B. curtipendula),* three-awns *(Aristida* spp.), hairy grama *(B. hirsuta),* and blue grama *(B. gracilis)* (p. 5).

Recovery of perennial grasses within the watershed varied both by rainfall and by individual management. "Almost the entire south end is under individual unit control. Many have taken some pains to protect the plant cover, and have good forage at present. Others have overgrazed their ranges severely and there is relatively poor forage now" (p. 8). "It is quite evident," the authors concluded, "that the entire area has been badly overgrazed in the past, but not particularly so at present, except in certain areas. Recovery can be established by protection and erosion control. The amount of recovery up to the present varies greatly. Generally there is very little recovery in the northern portion. . . . Such areas that have been protected in the past few years show indications of recovery by the presence of the more palatable perennial grass and browse species" (pp. 6–7). "Most of the southern portion which has a higher rainfall with quicker recovery from drouth conditions, etc., now supports a moderate cover of perennial grasses" (p. 9).

Unfortunately, the watershed report did not describe particular ranches, so conditions on the Buenos Aires must be inferred from other information. Clayton Vincent, who managed the ranch for the Gills from 1941 to 1955, recalls that the range was dominated by grama grasses and curly mesquite. There were only a few mesquites in the uplands, and the average herd on the ranch through the 1940s was about four thousand head, or twenty-five head per section. A reporter from the *Saturday Evening Post* offered this description in 1946: "[T]he sacaton and Johnson grasses grow rump-high in their time, and grama grass flourishes in the summer, while curly mesquite, alfilaria, Indian wheat and tallow weed provide a natural food reserve for the spring months of the year. [It is a] harshly beautiful, though almost treeless land" (Thruelsen 1946: 78). Accompanying photos show a large brushy mesquite tree growing in the embankment of a stock tank and an almost treeless plain in the background of a chuck wagon scene.

Overall, the Buenos Aires Ranch under the Gills was a produced landscape that produced beef. Like the draws and washes observed by Roskruge, spreader dams and stock tanks produced differentials of soil moisture that dramatically influenced vegetation. Under the altered regime, however, water was differently distributed than in Roskruge's time. Instead of infiltrating in place, more of it ran off into washes and then stock tanks, where it could be used by livestock. Evenly distributed water sources enabled even distribution of grazing pressure. In years of good rainfall the entire system apparently worked well enough that perennial grasses could recover relative to conditions in the 1920s. Calving rates ranged from 80 to 90 percent, suggesting that the cows were not short of feed.

This is not to say that the Buenos Aires Ranch was in equilibrium, however, or that it was as productive for wildlife as it was for cattle. Judging from later events, one can surmise that mesquites were growing in the old bottomlands by this time and would soon be identified as a problem throughout the ranch. In 1948 an AGFD wildlife biologist toured the Buenos Aires Ranch and found almost no wildlife (Knipe n.d.); a rancher who considered buying the ranch in 1952 recalls that it was dominated by brush and shrubby mesquites.

Triggered by drought, the same set of natural and artificial features that the Gills had established helped to provoke wholesale change from grassland to mesquite scrub. Unlike sacaton and Johnson grass, mesquites cannot tolerate inundation like the flooding that had occurred regularly in the bottomlands prior to arroyo formation. They require more water overall than perennial grasses, however, and once established their deep taproots allow them to survive droughts that kill more shallow-rooted species. After entrenchment, the old floodplain was thus a perfect habitat for mesquites, as were the earthen dams and stock tank embankments. Cattle lingering around water sources and eating mesquite pods helped to disseminate and fertilize seeds in the uplands. Especially in combination with grazing pressure, drought affected grasses much more severely than mesquites; this, in turn, assured that water would run off into tanks and dams when the rains returned, further aiding the trees. The soil "sheds water like concrete—every rain produces a series of flash floods along the arroyos," wrote the reporter for the *Saturday Evening*

Post (Thruelsen 1946: 80). The water-retention program thus depended on high runoff rates in the watershed while offering a partial antidote to their erosive consequences.

BREEDING

The reporter for the *Saturday Evening Post* characterized the Buenos Aires as a model modern ranch, "a strictly commercial enterprise consecrated to putting some hundreds of pounds of beef on an animal in the shortest possible time" (Thruelsen 1946: 26). Two years later, in 1948, the focus of production shifted from putting pounds on steers to producing calves. Although their stocker cattle had already been of "improved" breeds for years, the Gills now made breeding the critical productive moment of their Buenos Aires operation. They chose to bring in Hereford and Hereford/Brahma cows from another Gill ranch in Oregon. The decision was not unusual: Herefords were the dominant breed on western ranches for much of the twentieth century, and cow-calf operations had become the norm in southern Arizona by this time (Wagoner 1952). The history behind "improved" breeding requires examination, though, to understand the forces that shaped these developments.

It is necessary to start with some basic definitions. "Purebred" or "blooded" cattle are those whose ancestry has been controlled with an eye toward producing offspring with specific productive qualities (tender meat or greater milk production, for example). Because the genealogy of each animal is carefully documented, pure breeds are also known as "registered." "Conformity" refers to these animals' generally homogeneous physical characteristics (coat color, horns, size, and shape), resulting from generations of inbreeding and rigorous culling of nonconforming individuals. The major pure breeds of the late nineteenth century all came from Great Britain: Shorthorns in large numbers as early as 1834; Herefords after 1840; and Angus beginning in the 1870s (USDA 1900: 632ff.). In the twentieth century, Brahmas from south Asia gained adherents for their suitability in hot, arid environments. "Improved" breeding is the practice of crossing purebred animals (usually bulls) with unregistered stock. Ideally, the resulting offspring benefit from "hybrid vigor" while inheriting the productive qualities and conformity of the purebred sire. Pure-breed operations are a small minority of western ranches, especially in recent decades; the vast majority of ranch herds have been improved over several generations through crossbreeding.

According to Wagoner (1952: 38), most cattle in southern Arizona before 1880 were "Mexican breeds, handled in small herds by Mexican *rancheros.*" These animals had adapted to local conditions and were not "improved" in the sense just defined. William Oury is credited with being the first to import purebred cattle to southern Arizona in 1858; Colonel Henry Hooker and Colin Cameron invested heavily in Durham, Devon and Hereford cattle in the 1870s and 1880s (Morrisey 1950: 152–154). The practice spread quickly after 1880, and today virtually all the range cattle in Arizona (except some on Indian reservations) are the products of careful breeding for genetically determined traits.

From the cattle boom forward, the logic driving improved breeding was circular. The British, having "fixed" pure breeds through careful selection from the late eighteenth century on (Trow-Smith 1959), had come to expect the tender meat and uniform carcasses toward which breeders had set their sights. These tastes carried over to the eastern United States. As the beef industry increasingly relied on eastern and British markets for both investment capital and final sales, supply and demand alike intoned, "Improve the herd." John Clay Jr., a Scottish cattle merchant turned Wyoming rancher, described the process eloquently: "Undoubtedly, the foreign demand has been the greatest incentive to improvement. . . . We go to the parent country; buy in Aberdeen their best Shorthorns and Angus cattle; from Hereford and other parts of England we import the best White-faced blood. Streaming through our native pure-bred herds it reaches in diluted form our feed-yard steers, and then it returns across the ocean, giving that reciprocity of trade which England cultivates so generously" (USDA 1900: 631).

Thus was a socially constructed judgment of taste transformed into an economic imperative. The fact that improved animals fetched a higher price on the market was taken to reflect intrinsic superiority. In penalizing producers who did not or could not invest in these distinguished beasts, the market applied on a national scale the same principle that Robert Bakewell and other renowned breeders had applied to their herds:

"to cull drastically" (Trow-Smith 1959: 47). Controlled selection and breeding of livestock thus appeared as a "natural" response to market forces, and economic dominance as a function of better blood.

In Britain improved breeding had been largely a private-sector project, patronized by the upper classes; in contrast, U.S. government agencies such as the USDA's Bureau of Animal Industry constructed it as a democratic, public mission. The mechanism of improvement would be the market rather than the aristocracy, prices rather than principles: Good breeders would earn rewards to which others would aspire. The market could not compel improvement, however, until American producers understood the criteria for good breeding. Above all, they had to learn a particular way of seeing cattle. An early bulletin announced, "There is at the outset a well-defined beef type that admits of less flexibility than is generally supposed. . . . So clearly and definitely is this beef type established that to depart from it means to sacrifice beef excellence" (Curtiss 1898: 4–5). Using pictures and diagrams, it prescribed the use of a scorecard to appraise the beef animal against twenty-five attributes, differentially weighted and adding up to a possible one hundred points. The scorecard encouraged cattle buyers to imagine live animals as walking assemblages of butchered parts, each with a per-pound price. The trick was to discern the internal qualities of the animal's meat from external appearances, as if by using a sort of x-ray vision. Maximum profits defined the end; uniform criteria embodied in a common gaze defined the means.

The power of buyers to "educate" producers evolved in tandem with the division of labor in beef production and processing. Rather than raising calves through to adulthood and selling them at three or four years of age for immediate slaughter, as they had before, in the 1880s southwestern ranchers began to ship yearling cattle to Midwestern pastures for fattening, a practice that spread to Arizona during the great drought of 1891–93: "Until 1892 the generally accepted policy was to retain all she stock and sell range-grown three-year-old steers. At that time, however, *northern buyers refused to accept* the three's unless the two's could also be purchased. In 1890 the average age of marketed range cattle was 2.18 years. Ten years later it had been decreased to 1.63 years. The age has been lowered until now the raising of calves or yearlings is followed almost altogether" (Wagoner 1952: 45, emphasis added).

The shift to calf production had its roots in the ecological damage of the cattle boom and in unequal power relations between producers and middlemen. Yet both of these factors were obscured in contemporary rhetoric, which emphasized the benefits of a national division of labor and explained Arizona's position positively in terms of climate. Arizona's mild winters favored calving there over other regions, whereas its relatively sparse vegetation (after the boom) made fattening more difficult. As Americans grew accustomed to eating more beef, maximizing productivity through specialization was constructed as both patriotic (especially during wartime) and "natural." When Dr. R. H. Williams of the University of Arizona addressed the Arizona Cattle Growers' Association shortly after the outbreak of World War I, technical and ideological themes reinforced each other: "Beef is used in all parts of the world inhabited by civilized man. Its use is an index to wealth and intelligence of people. Thus, the per capita of beef is 70 pounds in the United States; 56 pounds in Great Britain; 39 pounds in France and 32 pounds in Germany." Arizona "is a natural nursery" for cattle, he explained, and cattlemen should obtain registered bulls and weed out "inferior stock." "A good slogan for every cattleman in this state would be 'Not more cattle but better cattle.'"[5]

Extension publications transmuted the drive to maximize profits into a narrative of progress through the manipulation of nature. In *The Range Bull*, published in 1925 by the University of Arizona Agricultural Extension Service, Charles Pickrell declared that "[w]e are a range cattle-breeding State" and that quality cattle were required in an era of rising costs of production. "The necessary improvement can best be made by introducing pure blood through the sire" (1925: 10). Pickrell explained that consumers preferred marbled meat and that marbled meat came from animals with "many more layers of tissue and a much greater space for the deposit of fat between these layers than is found in the carcass of the common or 'cold-blooded' animal. An animal is born with as many layers of tissue as it will ever have. Whether this number of layers is large or small depends upon inheritance." Fat on the outside of the meat is objectionable, but the consumer "is willing to pay even a premium for meat containing fat when it is placed in the form of marbling" (ibid.: 10–11). Pickrell illustrated his claims with photographs of round steaks and a microscopic view of meat tissues. Comparative photographic lineages underscored the point that greater conformity in ancestry made for greater conformity in offspring. Only a "registered" bull, whose ancestry was documented and certified by breeders' associations, could ensure such conformity.

By focusing on breeding as a "scientific" means for individual ranchers to strengthen their competitive position vis-à-vis other cattle producers, state agencies helped depoliticize latent issues surrounding the relation of all cattle producers to the beef industry as a whole.[6] Because maximizing profits was itself deemed "natural," the experts' counsel could appear as nothing more than scientific fact based in plant and animal biology. Yet the point at issue was not so much nature as the distribution of economic risk. Between the birth of a calf and the final sale of butchered meat, ownership changed hands several times: from rancher to stocker to feeder to slaughterhouse to butcher (with some owners filling two or more roles). Only the last could directly inspect the retail product prior to purchase. How were the others to judge the value of meat that was still alive in the form of a cow? Meatpackers—already a highly concentrated sector as early as the 1880s—had a vested interest not only in keeping prices low but also in obtaining uniform carcasses to smooth passage through mechanized, high-volume slaughterhouses and to yield standardized final products. Improved breeding euphemized the power of these interests over the everyday practices of ranchers. After all, uniformity in live animals was not in itself a guarantee of quality meat. Uniformity was rather a side effect of the practices that had given shape to the improved breeds. By threatening to bid low on "native" or "scrub" cattle—which meant, in effect, any cattle that were not uniform—buyers passed the risk back toward the ranchers who bred bulls to cows. If breeding was about dominating nature, as Knobloch (1996) and Lawrence (1982) argue, it presupposed another, unspoken domination: that of ranchers themselves by industrial beef processors, mediated by the "invisible hand" of the market.

THE DROUGHT OF THE 1950S AND THE MESQUITE MENACE

By the end of World War II, virtually every aspect of American beef production had been "improved" through the application of science, technology, and economies of scale in production, processing, and marketing.

For proponents of the modern, industrial "new model," it was as though the pastoral basis of livestock production had been completely subsumed by the logic of capital:

> "What happens," asks the 1948 *Annual Report* of Texas A. and M. College, "when good crossbred cattle are put on improved, fertilized pasture, given mineral supplement, sprayed with DDT for flies, treated with rotenone, phenothiazine, BHC and hexachloroethene, and rotated from pasture to pasture with the seasons? A good many Texas stockmen already know: Close to 100 per cent calf crop and calves up to 200 pounds heavier at weaning time!" In other words, modern methods make money. (Sonnichsen 1950: 8–9)

The postwar boom drove up prices, prompting expansion and attracting outside investors both small and large.

In the arid and semiarid West, however, industrial innovations could not overcome the dependence of range cattle producers on rainfall, forage conditions, and market prices. When the market slumped in the early 1950s, many cattlemen found themselves heavily in debt, and some blamed outside (i.e., nonrancher) speculators for inflating the bubble (*Arizona Daily Star,* 11 November 1953). High debt loads and low cattle prices encouraged ranchers to increase their herds still further, and cattle numbers in Arizona reached a thirty-year high at the end of 1955 (*Arizona Daily Star,* 26 February 1956). By the following summer, extreme drought conditions were forcing cattlemen to sell surplus animals early and to haul water to areas of the range where grass remained. The federal government extended disaster relief in the form of subsidized supplemental feed (*Arizona Daily Star,* 27 September 1956).

The drought had actually begun in 1951, and scientists have since determined that it "was one of the most severe in the Southwest during the last 350 years, and, perhaps, was exceeded in severity only by the Great Drought of 1275–99" (Herbel and Gibbens 1996: 2). Its effects on rangeland vegetation, though still not fully understood, were significant and enduring, even in areas excluded from grazing.[7] On the Buenos Aires Ranch, the drought appears to have upset the balance achieved in the 1940s, suppressing grasses and expanding opportunities for nonforage shrubs. By 1959, when Roy Gill sold the ranch, the herd numbered only two thousand head (*Arizona Daily Star,* 21 November 1959: B1).

It was in this context of ecological and economic malaise that the vision of intensive range restoration resurfaced. Research had continued on the Santa Rita Experimental Range for decades, but results were desultory; most attention had turned to breeding, feedlot fattening, water development, and disease and erosion control. Now, as though the very intransigence of the range posed a challenge that had to be met on principle, efforts were redoubled to restore earlier range conditions in the Southwest.

Half a century before, when Jared Smith and David Griffiths had conducted their experiments, the problem had been a lack of vegetation of any kind. This time, in contrast, there were specific botanical enemies: plants that had "invaded" the grassland and were outcompeting the grasses. Some were small but prolific shrubs such as burroweed and snakeweed; at higher elevations, piñon and juniper trees were increasingly dominant. The most notorious and daunting invader in southern Arizona, however, was the mesquite tree (*Prosopis* spp.; in the Altar Valley, *P. velutina),* which had been limited to drainages and floodplains at the time of settlement but now occupied large expanses of uplands. Instead of the tall, erect, and isolated trees that Roskruge described, these mesquites were growing in high densities of short, multiple-trunked, thicket-forming plants, deterring cattle even where grasses managed to grow beneath the low canopy.

Researchers at the Santa Rita Experimental Range had been documenting the spread of mesquites there ever since Griffiths noted the phenomenon in 1910. In 1945 they had initiated an experiment in which trees were manually removed (using prisoner-of-war labor) to different densities on four study plots, with a fifth plot (with 138 trees per acre) used as a control. They found that as few as twenty-five trees per acre diminished grass production by approximately 50 percent. This raised two questions: How to get rid of the trees, and how to replace them with some plant or plants that would provide forage for livestock. Researchers at the University of Arizona Agricultural Experiment Station had come up with two answers to the first question: grubbing out seedlings by hand and spraying the base of trees with diesel oil. But neither method was economical on dense stands of mature trees, so the second question was moot (UAAES 1959: 15–17; *Arizona Daily Star,* 1 April 1956).

In late 1955 the USDA announced a cooperative research project with the Arizona Agricultural Experiment Station, making available increased funding to tackle these two questions anew. Some researchers screened and tested native and non-native grass species in an attempt to develop hybrid strains for "a full-scale breeding program." Others conducted laboratory and field tests on a variety of chemicals that, it was hoped, would allow mesquites to be killed by aerial spraying. Bulldozers and other heavy equipment were used to remove mesquites and prepare the soil for seeding (*Arizona Daily Star,* 27 November 1955; Roundy and Biedenbender 1995; UAAES 1959). By 1957 more elaborate recommendations had been developed for eradicating mesquites: grubbing for small trees (less than one inch in diameter), diesel oil for medium-sized trees (one to three inches), chaining for trees over three inches, and aerial spraying with 2,4,5-T (trichlorophenoxyacetic acid) for dense stands of any size class (Reynolds and Tschirley 1963).

None of these methods was found to be economical on any large scale. Chained mesquites often resprouted from the root crown. Aerial spraying killed no more than half of the trees and required reapplication (UAAES 1959). Both chaining and reseeding suffered from high capital costs for heavy equipment (Roundy and Biedenbender 1995). Efforts to develop reseeding techniques took longer and were only moderately more successful. Lehmann lovegrass *(Eragrostis lehmanniana),* a South African perennial first imported to Arizona in 1932 (Anable et al. 1992), showed promise for erosion control and, to a lesser degree, for livestock forage. Two other non-native lovegrasses and their hybrids achieved similar results (Anable et al. 1992; Roundy and Biedenbender 1995; see chapter 6). Even the best species, however, were highly dependent on favorable rainfall and seedbed conditions for establishment. On the whole, David Griffiths's conclusions from 1910 were reaffirmed.

The effort to eradicate mesquites received strong support in the Tucson press, where metaphors of invasion, larceny, and war dominated. "Grass, or other usable forage, is wealth in the range lands," opined the *Arizona Daily Star;* mesquites, however, were "the major menace" in southern Arizona (28 November 1955). Less than three weeks later, the editorial page accused mesquite trees of "stealing water" from valuable forage plants (15 December 1955). As the drought dragged on, the rhetoric grew more heated and the government research received more coverage. The idea that mesquites were stealing profits from ranching was reinforced by "scientific" calculations: a pound of mesquite pods consumed four times as much water as a pound of grass; areas cleared of mesquite produced more than five times as much grass (by weight) as untreated areas. Frontier-era boosterism resurfaced in headlines such as "War On Mesquite Trees Gets Results—U.S. Tests Proving Conquest Possible." "Numerous ways of killing the water-greedy plant have been developed, but the search is still on for a more efficient and more economical method" (*Arizona Daily Star,* 1 April 1956). By the fall of 1956, the *Arizona Daily Star* editor had this to say: "Those well-meaning ones who protest that the juniper, the piñon and mesquite are beautiful to look upon and should be undisturbed should read with care the words of Ray Cowden, veteran Arizona range man. He says, 'Juniper is not something of value, but it is a parasite and if we do not control the spread of it, it will destroy us'" (19 September 1956).

In an atmosphere of public hysteria, several ecological insights were lost in the rush to develop mechanistic techniques of range improvement. Like Griffiths before, University of Arizona range ecologist Robert Humphrey (1958, 1962) concluded that fire suppression, more than any other single factor, was responsible for increasing mesquites, and he experimented with prescribed burning (*Arizona Daily Star,* 28 July 1956). Mesquite-infested desert ranges, however, generally lacked the necessary fine fuels to carry a fire, and public sentiment against range fires (like forest fires) was strong. When the USFS published the leaflet *Mesquite Control on Southwestern Rangeland,* no mention of fire as an ecological factor or as a management tool was included (Reynolds and Tschirley 1963). Meanwhile, range scientists continued to insist, as they had since Griffiths, that the techniques they were advocating could not work in the absence of basically sound grazing practices (UAAES 1959). But the dramatic and novel methods spoke more persuasively to the desire for radical change, and their greater cost could be written off as a capital expense. They were also more likely to be featured in publications, whose covers depicted bulldozers, people spraying diesel, and helicopters spraying chemicals. Humphrey asked soberly if Lehmann lovegrass might actually "finish the job" begun by mesquites—that of driving out native grasses (UAAES 1959: 28)—but almost all public land-management agencies, as well as highway departments, continued to use it for erosion control and range restoration purposes (Anable et al. 1992: 181).

Research into mesquite control continued into the 1970s, when it was applied on the Buenos Aires Ranch. I will examine its ecological effects in more detail in chapter 6. For now, it suffices to observe that the methods developed failed to meet the economic criterion for success: returns of sufficient magnitude and certainty to justify the expense. A recent review concludes that after decades of work, "these treatments do not ensure success. Instead, revegetation success is largely determined by the pattern of summer precipitation in a given year" (Roundy and Biedenbender 1995: 294). This did not prevent ranchers from applying the new methods, and if done under favorable circumstances (low costs, timely rainfall, and high cattle prices), they could pay for themselves. Such circumstances, however, were unpredictable and usually short lived.

CONCLUSION

Roy Gill dedicated much of his time to quarter horses. The most famous story from his tenure on the Buenos Aires Ranch recounts the day in 1947 when his horse Barbara B. won an unofficial race at Hollywood Park. Gill bet $50,000 that Barbara B. could beat any thoroughbred in a quarter-mile sprint; Charles Howard accepted the bet, putting forward his famous sprinter, Fair Truckle. Barbara B. won by two and a half lengths. Gill used his winnings to build a new horse barn on the Buenos Aires Ranch. In local lore, the glamour and audacity of "the $100,000 race" have over-shadowed most of the Gills' activities on the ranch itself (Leavengood 1993).

The legacy of Barbara B. testifies to the enduring grip of ranching's romantic appeal and to the blend of high society and high risk that continued to characterize range cattle production through the Gill era. "The rancher, like any financial speculator, thrives on the calculated risk. But the concept of risk implies a degree of indeterminacy in the workings of nature" (Ingold 1980: 284). Scientific research sought to minimize the risk by eliminating the indeterminacy; "the trend, in general, was reductionistic, more and more intensively or precisely pursuing the individual component pieces of resource problems" (Wasser 1977: 76). One wonders: If the research were ever completely successful, might ranching lose its romance? After all, it is hard to tell captivating stories about feedlots.

The Buenos Aires Ranch under the Gills walked this line. It was indeed a model modern ranch with fine cattle, an expensive infrastructure, and management that appears to have been up to date with the latest thinking in range science and extension. Native perennial grasses returned following the crisis of the early 1920s, in apparent confirmation of Sampson's theories of succession and grazing management. Lower stocking rates and range improvements may well have been responsible, at least in part, for a partial restoration of pre–cattle boom conditions. But risk remained in unintended consequences triggered by unpredictable rainfall. The standard model could not accommodate the severe drought of the 1950s, nor did it anticipate the encroachment of mesquites. Even without grazing, the drought might well have led to increased mesquites by dampening competition from grasses. But the redistribution of moisture caused by arroyo formation, stock tank construction, and grazing appears to have contributed to this complex shift in vegetation. In direct violation of the model, the shift would not reverse itself in the decades to come, despite further declines in stocking rates.

5

The Urbanization of Ranching

The decade of the 1960s was a relatively uneventful one on the Buenos Aires Ranch. The routine activities of managing cattle, mending fences, and maintaining water sources continued, but the ambitious improvement programs that characterized the La Osa and Gill periods fell into dormancy. While a few neighboring ranchers bulldozed mesquites, the two owners of the Buenos Aires, who bought and sold the ranch in relatively rapid succession, left without making much of a mark on the landscape. Mesquite trees consolidated their dominance of the vegetation, and the number of cattle on the ranch declined, despite attempts to increase the herd to somewhere near its size under the Gills. Beyond these basic developments, the environmental history of the Buenos Aires Ranch during this period is rather thin. This chapter explores the broader context of ranching in southern Arizona at this time in an attempt to provide some explanation for this abrupt decline in the intensity of management on the Buenos Aires Ranch.

The apparent lull should not obscure the importance of the changes that were occurring on a regional level during this period. In theory, the drought of the 1950s ought to have resulted in a flight of capital out of ranching similar to the one that had occurred in the 1890s. The value of ranches should have declined to reflect the diminished profitability of cattle production; foreclosures and bankruptcies might have swept through the area, as happened in the Midwest during the farm crisis of the 1980s. No such devaluation took place, however. On the contrary, ranch values climbed steeply from the end of the drought through the 1960s and 1970s. This was possible because a conjuncture of other factors operating on a national scale gradually shifted the basis of capital accumulation from cattle production to real estate development and speculation, overturning one of the fundamental premises of the formation of ranching: that rangelands "are, and probably always must be, of chief value for grazing" (Public Lands Commission 1905: 7). Well ahead of actual subdivision and home construction, ranches began to appear as potential suburban lands, at least in the eyes of finance capital. The events of the 1970s and 1980s, explored in subsequent chapters, must be understood in relation to the structural shift examined here.

Demographic and technological changes underlay the shift, but they do little to explain it in any detail. Growth in retirement communities, military and defense industry employment, and the urban sector in general increased demand for developable land. The expression of this demand in the landscape, however, depended on myriad extraeconomic factors. As two distinct regimes of accumulation collided on the range, their respective modes of regulation intersected (Hudson 2001), sometimes abetting each other and other times conflicting. Extension agencies adapted their expertise in grasses to serve new residents' landscaping needs. Where cattle wandered into suburban back yards, citizen outcry prompted changes in laws regarding fences, livestock, and liability. Policies and regulations ostensibly intended to encourage investment in cattle production now enticed outside capital into ranch properties, regardless of cattle prices, rainfall, or forage conditions. Changes in federal housing policies had a corresponding effect in stimulating demand for the final product: single-family homes underwritten by mortgages.

Curiously, the mythology of ranching did not disappear in the shift to an urban-dominated economy. It evolved into an ideology of suburban growth, a suite of symbols used in the development and promotion of "ranchettes": homes on the range, where the deer and the antelope play, etc. This is at once evidenced and occluded by linguistic practice. By one definition in the *Oxford English Dictionary,* a ranch is "a single-storey or split-level home." Many subdivisions, large and small, successful and unsuccessful, use "ranch" or "rancho" in their names. On the north side of Tucson is a campground called the RV Ranch. The Lazy S Auto Ranch is a used car lot. In Nevada countless legal brothels are called ranches. Donald Worster (1992: 37) notes with some sarcasm that "in California . . . everything is a ranch—out there they have diet ranches, avocado ranches, golf and tennis ranches, suburban ranchettes, and retired President ranches." It may be answered, of course, that these places are only called ranches, and that no one takes them to *be* ranches. But this begs the question of why people would choose such names for their homes or businesses. The point is that "ranch" is a title appropriated by different people to serve various ends. Without squabbling over what it means, they are using the symbolic value of ranching as a tool in a larger political and economic struggle. The newly dominant form of capital accumulation has legitimated itself partly by donning the clothes of its predecessor.

THE URBAN BOOM

During the Cold War, federal defense spending concentrated disproportionately in the Southwest, where the climate and vast open spaces allowed for year-round pilot training and weapons testing. Government spending on roads and highways also spiked upward, indirectly subsidizing widespread automobile ownership and use. Generalized postwar prosperity, especially among the middle classes, enabled significant increases in vacation travel and home ownership. The advent of air conditioning overcame the greatest natural deterrent to potential migrants from elsewhere in the country: oppressively hot summer temperatures.

People began to move into Arizona in droves. Between 1950 and 1960, the state's population increased 74 percent; in the following decade it grew another 36 percent, to 1.77 million people. The vast majority of these migrants took up residence in the Tucson and Phoenix metropolitan areas, home to almost three-quarters of all Arizonans in 1970 (*Arizona Daily Star,* 11 October 1972). The trend continues to the present day. As of 1997 the population stood at 4.55 million (a 257 percent increase from 1970), almost 80 percent of whom live in the two major metropolitan areas (map 3).

Economic and ecological differentials between Arizona and the rest of the United States reinforced each other in attracting these migrants. Especially for retirees from areas farther north, mild winter temperatures and low year-round humidity set Arizona apart. Relatively low land and housing costs meant that many migrants arrived with equity sufficient to buy larger properties than those they left behind. So much of the present population has arrived since World War II that in Arizona, urbanization *is* suburbanization, characterized by spatial expansiveness ("sprawl") and single-family homes. Land ownership patterns established during the formation of ranching compressed this new demand into a minority of the

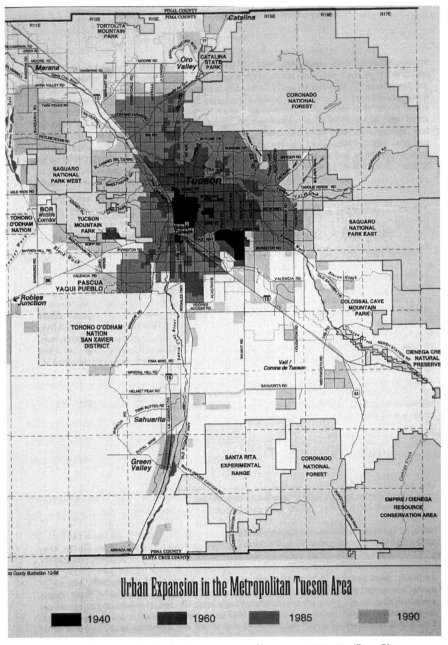

Map 3. Urban expansion in the Tucson metropolitan area, 1940–90. (From Pima County Board of Supervisors 2000, p. 10a)

state's overall acreage (deeded lands and a small portion of State Trust lands), causing ranch values to increase regardless of the economics of cattle production or ranch owners' activities. Today, the market value of a ranch is almost wholly determined by its real estate potential, not by the number of cattle it can support.

The drought of the 1950s was probably only a minor contributing factor to the overall dynamic of the urban boom, but it highlights the contrasting relations to nature embedded in the transition from ranching to real estate. Not only are homes vastly more lucrative per acre than cattle, but the capital invested in them is also relatively autonomous from Arizona's fickle climate. Most "new westerners" earn wages or draw pensions that are unaffected by drought. For their purposes, the desert environment is static: sunny and warm every year, whether it rains fifteen inches or five. Their baseline is not temporal but spatial: They compare Arizona's weather to that of New England or Iowa rather than to the weather of years past. This temporal security is precisely what the capital invested in ranching had sought in vain for decades; it embodies the same abstraction from time contained in contemporary government range-management policies and the concept of carrying capacity.

STATE AGENCIES AND THE LAWN

In the 1950s, while ranchers and scientists were declaring war on mesquites, newly arrived city dwellers embarked on a still more intensive and, by virtue of its scale, more attainable program of remaking the desert to fit ideals imported from the humid East. At the center of this program was the lawn.

As early as the 1890s, the Division of Agrostology had recognized "lawns and lawn making" as subjects that could serve a broader public constituency. It was in part a reaction to growing urbanization in the industrial regions. "Nothing is more beautiful than a well-kept lawn, whether it be of large or small extent. Even the small plots fronting city dwellings are points of attraction when covered with a soft, even turf" (USDA 1898: 355). The normative model was borrowed from the estates of the New England elite (who had themselves taken it from the English aristocracy and high bourgeoisie), but the USDA translated it into neutral, quasi-scientific terms for middle-class Americans. "The perfect lawn consists of the growth of a single variety of grass with a smooth, even surface, uniform color, and an elastic turf which has become, through constant care, so fine and so close in texture as to exclude weeds, which, appearing, should be at once removed" (USDA 1898: 367). More than science was involved, however. Bulletins and circulars preached that overcoming nature in one's yard required constancy and devotion; it was as much a matter of self-discipline as of technical expertise. "[T]here is nothing which more strongly bespeaks the character of the owner than the treatment and adornment of the lawns upon his place. . . . A perfect lawn can not be made in a season, and the highest excellence sought comes only through intelligent care for a period of years" (USDA 1898: 355). The lawn resembled a Protestant test of faith, individualized and private but also invidious and conspicuous.

Post–World War II migrants brought their notions of the proper lawn to Arizona, and government agricultural extension agencies responded with bulletins on domesticating the desert.[1] No less than in range research, confidence was high that the landscape could be forcibly transformed to suit the purposes at hand. *Lawns for Arizona*, a circular published by the University of Arizona Agricultural Experiment Station in 1951, opened with a photo of a trim suburban house with fake shutters, a neatly pruned hedge, and a thick lawn, all shaded by a tall arching tree. "Interest in beautifying home grounds is increasing throughout the state," it began. "The lawn is considered a starting point in planning an ideal setting for shrubs, flowers and trees. Lawns form a soft carpet for recreational activities. They also serve a useful purpose in reducing the amount of dust and mud that inevitably finds its way into the house. No matter how modest your home may be, a well-kept lawn gives the property a look of tidy spaciousness."

The private, single-family home was not the only domain where turf was viewed as necessary. Tourist areas, schools, parks, and ball fields also required lawns; the desert was simply incompatible with properly organized Anglo civilization. Through landscaping, the unpleasant, unfamiliar, or dangerous aspects of the desert environment could be economically altered or effaced, replaced with green grass and shade trees. And with reliable uniformity over time: Instead of "a dust bowl at some seasons and a pond at others," a well-kept park lawn would be "a year-around attraction to campus and town."

FENCES AND CATTLE IN SUBURBIA

According to extension pamphlets, the ideal lawn pushed the desert back from the single-family home, creating a domesticated buffer against rattlesnakes, cacti, dust, and scorpions. Domestic livestock were no more welcome than venomous wild creatures, however. The lease systems described in chapter 3 had ended the open range of the cattle boom, but another type of open range—one to which yellow road signs alert motorists to this day—endured. Under longstanding state and federal law, livestock in the rural West has a presumptive right to graze anywhere. Private landowners may fence them out, but livestock owners are not required to fence them in. When a domestic animal does damage to private property, liability turns on the issue of a "legal fence." (Such a fence must meet specifications as to number of wires, spacing of posts, etc.) If the land is legally fenced, then the livestock owner is liable; if not, the aggrieved party has no recourse under law (cf. Ellickson 1991). In cities, however, presumption is the other way around. In 1936, two years after the Taylor Grazing Act, the Arizona legislature gave itself the authority to designate "no-fence districts" in urban areas, where livestock had to be fenced in by their owners. As Tucson grew, its no-fence district expanded (eight times between 1936 and 1960, according to the *Arizona Daily Star*). Counties without large urban areas were sometimes called "cow counties" because of the ubiquity of livestock.

The rise of suburbia challenged this simple system.[2] Suburban properties were large by urban standards; fencing them was expensive and usually considered an eyesore. Moreover, the suburban ideal was neither urban nor rural: "The attraction of suburbia, from its origins in late-eighteenth century England, had been the chance for the middle class to flee high-density urban living, the 'corruption' of the city, and the constant press of contact with the city's heterogeneous population. Suburbs catered to an arcadian fantasy in which the nuclear family would be plucked from the social fabric of the city and placed in purifying contact with nature" (McKenzie 1994: 81). Cattle had no place in this suburban fantasy, at least not in postwar Arizona. They were present on adjacent rangelands, however, and were attracted by lush, well-watered landscaping, especially during dry seasons when natural vegetation was sparse.

By the end of the 1950s, suburban development was surpassing existing no-fence districts faster than the legislature could be moved to expand their boundaries. Many suburban residents, moreover, did not want to be incorporated into Tucson or Phoenix city limits. In 1960 State Senator David Wine introduced legislation to streamline the process of establishing no-fence districts in suburban areas of Pima and Maricopa Counties. He had failed in earlier efforts, but this time he exempted counties with fewer than 250,000 residents in order to placate rural legislators. Wine's bill transferred jurisdiction from the state legislature in Phoenix to county boards of supervisors, authorizing them to create no-fence districts upon petition by more than 50 percent of real property taxpayers in the affected areas. The *Arizona Daily Star* (18 February 1960) supported the bill, calling roaming cattle "an expensive and dangerous nuisance" and asserting that "[o]nly the careless and shiftless would knowingly permit their livestock to roam, untended and unheeded, on other people's property." Four years later, when the legislation finally passed, the editor attributed resistance to no-fence districts at the state level to the presence of "a lot of cowboys" in the legislature (23 May 1964).

Under the new law, suburban residents were empowered to launch petition drives to have their neighborhoods designated as no-fence districts. In 1964 retired Navy Admiral Edward Halloran, president of the Windsor Park Association on the eastern edge of Tucson, spearheaded drives for his own neighborhood and for an area north of town. As news of the efforts spread, stories of cattle in suburbia poured in to Halloran and the newspapers. Emphasis repeatedly fell on the dollar value of damage done by cattle to private property. Just as ranchers and earlier settlers had sought state assistance in protecting their livestock from mountain lions and wolves, suburbanites expected the law to protect their landscaping from the "depredations" of "marauding cattle." In two months, Halloran gathered twenty-three hundred signatures, representing 85 percent of residents in the east side area, which encompassed some seventy square miles (a density of fewer than thirty-nine people per square mile, or sixteen acres per person). The *Arizona Daily Star* again editorialized in support, and the Pima County Board of Supervisors declared the no-fence district on 21 June 1964. The north side area—some thirty-six square miles—was declared a no-fence district a few months later.

The first test of the new legal landscape was a decidedly unequal affair. Calling themselves the Pima County Livestock Association, seven ranchers filed suit in Superior Court, contending that the no-fence district was unconstitutional on the grounds that the petitioners, not the supervisors, had written the legislation. Within a week, the suit was discredited. According to county records, only two of the plaintiffs, Alfredo Campos and Gilberto Miranda, owned any land in the no-fence district; this, rather than historical use, was the basis of legitimate property in the new social world of the suburbs. The Pima County Livestock Association, moreover, did not exist as a legal corporation. The plaintiffs, it seems, were small operators, at least three of whom were of Mexican descent; they lacked the capital for leases, fences, and improvements. The *Arizona Daily Star* reported that they "have been feeding their cattle on trees and plants planted by homeowners. 'Where do you think they get their water?' asked the sheriff. 'In private fish ponds. And we even have a photograph of a cow wading in a private swimming pool'" (1 July 1964).

Alfredo Campos's predicament is particularly revealing. He owned less than two acres in the Indian Hills area, where adjacent land had been subdivided and developed as Indian Hills Estates in the late 1950s. In 1960 Campos had been assessed a two-hundred-dollar fine after a new resident filed a complaint against him. "Residents of the exclusive development have complained that loose cattle have damaged their shrubs," the *Arizona Daily Star* reported. Campos argued that he had been grazing his cattle there for years and that state law exempted him from liability. By the time of the 1964 trial, the law had changed but the story was the same: "Testimony . . . revealed that Campos' cattle wandered from the owner's $1\frac{1}{2}$-acre tract onto at least four private residences. The cattle caused a total of between $350 and $425 damage to shrubbery and grass, according to testimony by four eastside residents. . . . Rear Adm. Edward R. Halloran (USN-Ret.) . . . contended that the cattle ate his choice Burbank cacti, and Louis A. Wilson . . . said the livestock tore shrubbery and flowers 'loose from their moorings'" (*Arizona Daily Star*, 27 August 1964).

The Southern Arizona Cattlemen's Protective Association (SACPA), representing larger and mostly Anglo ranchers, came out in favor of the no-fence districts, seeking to preserve "the good name of the cattle industry in its relationship with the general public." Its members had enough land for their animals and enough capital to fence their pastures; they would not feel the pressure of suburbanization for several more years.

SUBURBANIZATION AND THE RENT GAP

Just as free land propelled the cattle boom of the late nineteenth century, the differential in land values for cattle production and for residential development helped drive the suburban boom of the post–World War II period. As elsewhere (witness Brazil today), ranching in southern Arizona served both the annual production of livestock and the historical production of commodified land: it transformed "raw" land into a form of private property that could be bought and sold on the open market. Market pressures could then influence land use through rent, expressed in land values, credit, appraisals, and tax assessments. In the absence of a rival land use, divided access to land was the keystone of responsible ranching because it gave each rancher incentive to conserve the range. Once a more lucrative use emerged, however, it became ranching's Achilles' heel, both economically and ecologically.

Building on the work of Henri Lefebvre and David Harvey, Neil Smith has isolated a particular feature of capitalist urbanization, which he terms "the rent gap" (1996: 51–74). Smith derives the gap from empirical work on gentrification in inner-city neighborhoods, a process that follows on the heels of extensive suburbanization. Properties in core neighborhoods, relatively deprived of investment capital during the period of suburban expansion, gradually decline in value. Buildings deteriorate, rents fall, and capital is devalued. A point is reached, however, at which actual rents are so low that it is worth the landlords' money to rehabilitate the property in search of higher rents. The exact point is defined by the rent gap: that is, the difference between actual rent and potential rent, the latter determined by larger social conditions and circumstances (supply and demand, relative location, state policies, etc.).

The rent gap is particularly conspicuous in the dynamics of urban gentrification, but its effects are by no means limited to the postsuburban inner city. Potential rent and actual rent diverge on the suburban fringe as well; indeed, I would argue that suburbanization is an even more striking instance of the rent gap than is gentrification. The disparity between agricultural and residential land values is not simply a qualitative difference between "raw" land on the one hand and "improved" land on the other. As we have seen, rangelands have had significant sums of capital invested in them during this century, and agricultural lands still more (especially in irrigated districts). We are dealing here with two types of capitalized land; the rent gap has its origins, as with gentrification, in devaluation coupled with larger social changes.

The urbanization of ranching consists in the historical process by which ranch lands, previously valued according to their capacity to produce cattle, have come to be valued according to their *potential* for residential real estate. Following Harvey and Smith, we can isolate two aspects of this process: the diminishing value-productivity of cattle (actual rent) and the increasing value of "undeveloped" land for housing (potential rent). At present, property values have increased to the point that buying a ranch for the purpose of cattle production makes little or no economic sense. According to several experienced cattlemen, a good ranch in Arizona today can cover a mortgage of forty-five to fifty dollars per acre for land, whereas even remote properties in southern Arizona cost two hundred dollars per acre and up.

Based on newspaper accounts, economic studies, and the testimony of long-time ranchers, it appears that this rent gap has characterized the region as a whole (rather than only land on the periphery of urban areas) since about 1970.[3] Venture capital outfits, drawn to the gap between actual rent and potential rent, expanded it further by driving up ranch values through their speculative investments. In the case of the fifty-five-thousand–acre Rio Rico development, for example, "General Acceptance Corp.'s predecessor, Gulf American Corp., bought the Baca Float ranch for a reported $3.5 million, or about sixty-four dollars an acre. GAC is now selling an acre for an average of three thousand dollars. Like the Baca Float ranch, most Arizona developments are on what was once considered prime grazing land" (*Arizona Daily Star*, 26 December 1971). In 1972 Robert Carpenter, a University of Arizona professor of urban planning, identified fifty-five land-conversion projects of a thousand acres or more underway or in preparation, covering some 643,000 acres of land. "The vast majority of the projects," he observed, "are programmed for lot sales in a speculative land market that is nationwide" (11 October 1972). In effect, real estate developers had already made arrangements for all the migrants expected to arrive during the following twenty years (26 December 1971).

Subdivision was a risky business, but the potential returns were irresistibly high. The costs of providing infrastructure—access roads, water, electricity, telephone service, and sewers—were enormous. Many developers attempted to sell lots without contractually obligating themselves to a time line for installing these services; they hoped to generate the needed cash flow through initial sales. This led to sleight-of-hand and hard, sometimes fraudulent sales tactics. If buyers did not materialize in sufficient numbers, the developers simply abandoned the project, leaving those who had bought parcels stranded. (GAC had done exactly this in Florida and had been sued.) Rio Rico stagnated for two decades, its roads ending in empty lots. Diamond Bell Ranch Estates, another GAC venture not far north of the Buenos Aires Ranch, is to this day mostly vacant, with roads badly eroded and street signs perforated with bullet holes. Only with passage of the Groundwater Management Act in 1980 were developers effectively curtailed from fraudulent sales practices.

The rent gap was not the only factor attracting capital into ranches, though it was prominent in any calculation of potential profits. Owning a ranch also had important tax implications, the result of policies crafted to encourage investment in ranches and ranch improvements. The value of improvements—which, excluding development potential, often represented the majority of a ranch's market value—could be amortized as a capital investment. A fence that lasted twenty years might be amortized three times, if each owner sold when the seven-year depreciation schedule ended. Moreover, capital gains in real estate could be deferred if they were quickly reinvested in another real property. Property taxes per acre were (and remain) much lower for agricultural land than for residential or commercial properties, or even for undeveloped "raw" land. Losses from a ranch could be used to write off profits from other economic activities.

Combined with the rent gap and high marginal income-tax rates in the uppermost brackets, these policies made ranching into a haven for capital: a tax shelter with the potential to yield a negative tax (Hymel 1998). Until the tax code revisions of 1986, it was economically rational to lose money in cattle ranching if one wished to shelter other profits from income taxation. Wealthy individuals and corporations flocked to invest their capital in ranches because much of the initial face value could be amortized even as the market value of the land was appreciating at well above the rate of inflation. From the drought of the 1950s forward, ranch sales became regular news items in Tucson's papers. By 1972 the two dozen largest landowners in Arizona owned 20 percent of the state's deeded acres; the twelve largest (a mix of land and cattle companies, insurance companies, mining companies, and developers) owned 1.85 million of the eleven million deeded acres in the state (*Arizona Daily Star,* 24 September 1972).

Most new owners continued to run cattle for tax reasons, and many pretended or aspired to be ranchers, participating in the symbolic regalia of hats, boots, and pickup trucks. "Instead of [a] typical second- or third-generation rancher, you are beginning to get a 60-year-old guy who is retired or who has outside income and is attracted to the ranching life," said one rancher (*Arizona Daily Star,* 7 January 1983). The *Arizona Daily Star* headlined its 1972 investigation "Arizona's Land Barons—Tall in the Saddle . . . of Their Cadillacs." In 1981 to the north of the Altar Valley,

> Ronald D. Cohn, a part owner of Super City Stores Inc. . . . took control of about 15,000 acres of ranch land in the valley, 14,000 of which is grazing land owned by the government and leased to him, from Oscar B. Robles. . . . "I think the land has a lot of great potential," [Cohn] said. "I think it'll only be five to 10 years at the most before people really start moving out there." Cohn said the land . . . attracted him not only as a speculative investment but also because of tax advantages and what he called a "lifelong dream" of owning a working cattle ranch. "It has 130 head of cattle on it right now and I want to keep it a working ranch for the time being," he said. "But there's no question in my mind that the Avra Valley will fill up with people. I'd like to put together a decent residential project,

ranchettes maybe. . . . It will be a nice setting for people who value a rural lifestyle, and it's only a 30-minute drive from the corner of Ina and Oracle roads." (*Tucson Citizen*, 21 March 1981)

These superficial continuities masked the extent to which suburbanization was undermining the economic viability of range cattle production. Outside investors were a mixed blessing. On the one hand, they brought needed capital into ranching, propping up equity values and facilitating credit. As one observer put it, they were "becoming bankers of the industry" (*Arizona Daily Star*, 7 January 1983). On the other hand, while speculation drove land values up—by as much as 100 percent per year at the height of the 1979–80 bubble (*Arizona Daily Star*, 1 February 1980)—the market price of cattle stagnated and operating costs climbed. The land eventually represented an enlarged mass of capital that could not be reproduced through livestock production. As early as 1966, net returns to capital and management from ranching in the West ranged "from very low to negative in all areas studied" (Martin and Jefferies 1966: 233; cf. Smith and Martin 1972). A 1983 study by the University of Arizona Extension Service found that an average three-hundred-cow ranch in Arizona was worth at least $500,000. Annual gross income was $51,000 and operating costs were $45,515, leaving an income of $5,485: only 1 percent on investment (*Arizona Daily Star*, 20 November 1983; cf. Fowler and Gray 1988). Many ranchers today characterize their situation as "land rich and money poor." Writing of the West as a whole, Paul Starrs (1998: 71) summarizes:

> Who, then, turns a sizable profit from livestock ranching? The answer is, hardly anyone. Studies of ranch economics in the 1970s turned to the techniques of cost-benefit analysis only to reach a startling conclusion, replicated in one study after another. At any size, on virtually any combination of private, leased, or loaned land, with cow/calf-raising operations or with outfits raising steers on grass, in virtually any sort of livestock venture, costs were likely to overshadow profits. By the early 1980s many range and agricultural economists were moving to concede the inevitable: Ranchers could not stay in business for purely financial returns.

As during the cattle boom, debt and market prices fundamentally structure the options of ranchers,[4] but now the price of cattle relative to the price of land is the crucial factor. Banks show an increasingly strong preference for real estate loans over ranch loans: Not only are returns per acre much higher, but with one of the nation's fastest-growing populations, Arizona's real estate market is robust. Moreover, newcomers' ability to pay off their mortgages does not depend on rainfall or vegetation growth for its success. The "New West" economy presupposes a built environment (or "second nature") that decouples growth from the environment, if only in the short term.

H. CLIFFORD DOBSON AND JOHN NORTON

Late in 1959 Roy Gill took his prized quarter horses and moved to the Lightning A Ranch near Tucson (*Arizona Daily Star,* 21 November 1959). He sold the Buenos Aires Ranch and its two thousand cattle to H. Clifford Dobson for a reported $2 million. Dobson promptly increased the herd to thirty-two hundred head. In 1962 he bought the Pozo Nuevo Ranch on the north end, bringing the total acreage of the Buenos Aires Ranch to about 116,000 acres. (The stocking rate was thus seventeen head per section.) The Dobson family had made their fortune in the Mesa area, first in irrigated agriculture and then in the conversion of farm fields into suburban real estate. The Buenos Aires Ranch appears to have been largely a speculative investment and tax shelter for them, based on the absence of major improvements during their tenure and the fact that they sold after seven years, the moment at which amortization of the ranch's existing improvements was complete.

John Norton and his wife appear to have had a similar interest in the Buenos Aires Ranch during their ownership from 1966 to 1972. Norton's company produced lettuce in the irrigated fields of the Yuma area. Their lasting marks on the ranch were unrelated to grazing. In a campaign of architectural normalization, they completely renovated Aguirre's old adobe ranch house. The old windows (which "raised into the attic on ropes") were replaced with steel sash units; carpeting, insulation, sliding glass doors, a heat pump, and new wiring were installed. In Mrs. Norton's words, "The plastered walls were so crude and crooked that it took 6,000 bags of cement to straighten the inside and outside walls of the house."

Outside they installed lights on lampposts to illuminate the driveway, and they built an airstrip "long enough for relatively large planes to land. . . . We communicated our arrival by flying over Headquarters, then the foreman would drive out to the airstrip and pick us up." Finally, they converted the original bed of Aguirre Lake into a pecan orchard, which subsequently flooded, killing all but a few of the trees. "When we awakened in the morning and looked out the windows we saw white cranes perched on the top of each tree. It was an amazing sight" (Leavengood 1993: 25–26). When the Nortons sold the ranch in 1972, it came with twenty-two hundred cattle on paper; according to the buyer, only sixteen hundred were there.

THE SUBURBAN MYTHOLOGY OF RANCHING
In chapter 3, I argued that the mythology of ranching isolated and romanticized pastoralist aspects of the cattle boom rather than those of ranching proper. The vigorous, rugged individualist ethos of the cowboy provided a counterpoint to urban, workaday life constrained by social propriety and the industrial clock. The idea that wealth derived from nature rather than labor appealed to aristocratic yearnings and to the antilabor sentiments of corporate industrialists. Attributes of the landscape—expansive vistas, starry skies, and abundant wildlife—came to define a lifestyle constructed in opposition to urban industrial capitalist society. Ranching represented a valorized "Other"; ranches were places where people lived a qualitatively different life.

These ascribed virtues of the ranching lifestyle are prominent features in the promotion of residential subdivisions today. Many are called "ranches," sometimes retaining the names of the cattle operations they displace. A newspaper advertisement for the "Grand Opening" of a new subdivision east of Tucson is representative:

Escalante is the newest offering of the incomparable Dragoon Mountain Ranch—a low density gated community of approximately 400 ranch sites set amid a sprawling 18,000 acre reserve. The ranchland at this mild 4,400' elev. is diverse and alluring; lush foothills vegetation, gently terracing hills, rich meadows (perfect for horse grazing), rich soils, abundant pure water and equestrian trails complete the picture. The view from the ranch is spectacular; 4 wilderness mountain ranges surround the Ranch, with the spectacular 52,000 acre Coronado National Forest hugging the Ranch. Wildlife is abundant and exotic. The nearby San Pedro River attracts over 300 species of birds to the area. Deer, fox, coatimundi, blue heron and others are frequent visitors. Build your dream-vacation-retirement-retreat at Escalante where your lifestyle is protected for now and for the future. Intelligent covenants (no parcel splitting, no mobiles) protect your lifestyle. Underground power service protects your views.

The class bias of this advertisement should be obvious, but also note the implicit ecological claims. Environmentalists routinely assert that arid and semiarid rangelands have been permanently degraded by cattle grazing, to the detriment of wildlife. Yet the moment the cattle are removed, the same land can be described as a paradise of scenic beauty and abundant wildlife. It is Clementsian ecological theory appropriated by professional marketing. The descriptions are exaggerated, to be sure; one could argue that few people take them literally. That is not the point, however. The advertisements appeal to a ranch fantasy, and fantasies are powerful for reasons other than their literal truth value. The question is how the features and attributes of ranching have come to represent an antithetical land use, one that eviscerates the very qualities—solitude, wilderness, wildlife, and "nature"—that it appropriates to promote itself.

The suburban mythology of ranching has emerged from a conjuncture of political and economic forces described in Evan McKenzie's (1994) social history of Sunbelt subdivisions. McKenzie explains how the common-interest development (CID) became the dominant form of suburban residential land use after 1960. As suburban development in California and Florida accelerated during the postwar period, the supply of land in those places began to shrink and land prices escalated, threatening the continued viability of the suburban model (McKenzie 1994: 80). CIDs were a reaction to this constraint, though they were subsequently applied in areas, such as Arizona, where the land shortage was not yet acute. The CID model allowed developers to increase housing densities by setting aside "common" areas (golf courses, parks, recreational facilities, and unbuildable areas such as washes) and then calculating per-house acreage for the development as a whole rather than for each lot. Once the Federal Housing Administration signed off on this innovation in 1963, extending its mortgage insurance to CIDs, the potential profits of large-scale land conversion rose significantly. Local zoning codes were quickly modified under intense pressure from builders to capitalize on the new opportunity.[5]

The problem with the CID model was that its higher densities threatened to undermine the "arcadian utopia" to which suburbia aspired. Some new way of promoting the suburban lifestyle had to be found. Negatively, the putative shortcomings of living in central cities were reiterated, ever more loudly as the Civil Rights movement advanced. Positively, the keystone of CID promotion became equity: "Above all, NAREB [the National Association of Real Estate Boards] endeavored to convince the public that buying real estate was a safe, conservative investment. . . . To expand private home ownership it was necessary to establish a sense of security about property values. The industry used its resources and influence to eliminate or minimize factors that appeared to make the value of residential property unstable" (McKenzie 1994: 61). Marketing studies suggested that planned communities were seen "largely as a way to safeguard property values" (ibid.: 97).

A sociospatial differentiation ensued, as communities were targeted to specific demographic groups—retirees, golfers, equestrian enthusiasts, etc.—or socioeconomic strata. When explicit racial covenants were ruled unconstitutional, "homeowner associations and restrictive covenants shifted their emphasis to class discrimination, which is legal, from race discrimination, which is not" (McKenzie 1994: 78). Covenants and deed restrictions evolved to enforce standards of physical appearance: "Unkempt yards, peeling paint, and other indications of neglect that might affect the neighborhood's property values are subject to censure" (ibid.: 129). Developers formed home-owners' associations to enforce these rules (usually backed by the threat of liens) and to manage the community-owned "public" areas within the developments. As McKenzie documents at length, CIDs embody and promote a "unique idea of citizenship . . . in which one's duties consist of satisfying one's obligations to private property" (ibid.: 196).

In the name of property values, the suburban demonization of the city has expanded to include anything unpredictable or potentially offensive—a house painted an unusual color, an RV parked in a driveway, or a grandchild living in an "adult" community—on the grounds that they might diminish the marketability of other properties in the community. "CID residents, in their concern with property values, often behave like the stockholders in any large corporation, who neither know nor care about corporate affairs as long as their stock goes up and they keep receiving dividend checks" (McKenzie 1994: 142–143).

McKenzie's characterization of home-owner associations may find detractors, but his insights regarding property values shed light on the suburban mythology of ranching in the late twentieth century. The pastoral elements mythologized a century ago have been transposed to fit a new context. With residential property values steadily climbing for the past forty years, developers and home owners alike (not to mention civic officials, bankers, insurance companies, and related professional-service providers) have come to view equity as a "natural" source of wealth, one that grows by itself. The mythical "lifestyle" of ranching has been decoupled from livestock and hitched onto real estate underwritten by home-owners' debt. Cattle now appear as symbols not of self-reproducing wealth but of toil and degradation: Dirty and domesticated, they desecrate the ersatz "nature" on which equity values are perched. Only the "amenity values" of the land—contrived isolation, expansive views, and wildlife (safe wildlife such as deer and quail rather than wolves or bears)— remain to simulate the mythical ranching experience of the last century.[6]

CONCLUSION

With the population boom of the post–World War II period, the urbanization of ranching in southern Arizona has proceeded simultaneously on economic, ecological, and sociocultural fronts. Economically, land values have increased to the point that cattle can no longer pay off ranch mortgages even in rural areas. This, in turn, has undermined the structural condition on which ecological stewardship in ranching must depend: long-term tenure on the land. The prospect that one's ranch will become a subdivision sooner or later, and that for such a purpose the health of the range is irrelevant, makes it rational to abandon long-term stewardship in favor of shorter-term profits. Thus the economic pressures of urbanization have potential environmental consequences even before ranches are subdivided.

Ecologically, urbanization poses a threat to the desert that is at once less visible and more critical than grasses and vegetation: water supplies for a population far larger than the area has ever sustained. The depth to water in wells in the Tucson Basin has increased steadily since the 1930s, and groundwater pumping today is approximately nine times what it was in 1930 (Tellman et al. 1997: 21–22). The arrival of the Central Arizona Project—a massive canal from the Colorado River to Tucson via Phoenix—only postpones the problem while reaffirming the power of state-backed capital to produce a landscape for further accumulation.

Politically, the real estate boom illustrates, as did the cattle boom before, that the most lucrative lines of economic activity typically elude effective government regulation until the natural bounty that they exploit has been exhausted. Socioculturally, the pastoralist ambition to profit from "natural" bounty has metamorphosed into suburban expectations of rising property values. Home owners willingly shoulder huge debts in pursuit of "great views," the "right" neighborhood, or "the ranching lifestyle." Finance capital, this time in the form of mortgages, underwrites wholesale conversion of the desert into a habitat for leisure.

Viewed historically, the urban boom recapitulates the cattle boom, but with humans instead of livestock. Outside capital has flooded the region, this time in the form of mortgages and pensions flowing through national real-estate markets and financial systems. As before, the capital is drawn to "natural" values that are so abundant as to be nearly free: sunshine, mild temperatures, picture-postcard scenery, and the (subsidized) recreational opportunities afforded by large national forests and parks. The ecological consequences of the cattle boom were severe and long lasting, but they will likely pale in comparison to those resulting from the late-twentieth-century real-estate boom.

Both booms illustrate one fundamental point: Capital may be likened to an ecological force. It can invade and adapt to an area as rapidly and opportunistically as any non-native species, taking advantage of any bounty "spontaneously provided by nature." After the deluge, some capitals will attempt to reconstruct the conditions for accumulation on the same basis as the initial free-for-all. If another, more lucrative or less uncertain form of accumulation emerges, however, one that capitalizes on attributes of the natural environment still readily available, then a process of transformation will ensue. It is no coincidence, in this light, that ranching became the target of environmentalists' wrath not when the damage was at its worst but generations later, when ranchlands began to be attractive investments for subdivision.

6

Restoring Grass and Quail

In March 1972 the Nortons sold the Buenos Aires Ranch to the partnership of Wayne Pruett and Peter Wray for $3.5 million. With the acquisition of two adjacent properties shortly thereafter, Pruett and Wray brought the size of the ranch to 182 square miles. In 1974 Pruett and Wray reorganized their ranches as the Victorio Land and Cattle Company, a subsidiary of the Victorio Company. In the history of the Buenos Aires, only the cattle boom saw more dramatic change than the Victorio era, and no period is more misunderstood.

Two parallel programs of ecological restoration unfolded in the 1970s, one conducted by Pruett and Wray and the other by the FWS in cooperation with the AGFD. The Victorio Company worked to restore the ranch from mesquite scrub to grassland conditions, primarily by bulldozing and reseeding. Concurrently, wildlife biologists worked to restore the masked bobwhite by releasing captive-bred birds. Although only loosely coordinated, the two efforts shared a common vision of what the Buenos Aires Ranch should be: a landscape dominated by grasses, with mesquites contained to the margins of drainages. This would benefit cattle by increasing forage and presumably help the masked bobwhite as well by returning the vegetation to its pre–cattle boom conditions. Over time, the two programs converged on the landscape, as each recognized the loamy bottomlands and adjacent alluvial uplands as the most promising areas for their respective ambitions. This chapter examines and contrasts the methods, successes, and limitations of these two programs, out of which the Buenos Aires Refuge was to be born.

The motives behind restoration were more complex than it might seem. The ranch had all the markers of prosperity. Through cutting-edge work in genetics and breeding, Pruett and Wray developed a world-class pure-breed operation specializing in Hereford and Brangus cattle. Buyers from North and South America landed private jets at the Buenos Aires airstrip to attend auctions; a one-third interest in one bull sold for $335,000. The cost of mesquite removal, seeding, fence construction, water improvements, and erosion control was around $1.5 million. Yet in every year but one, the Victorio Company lost money on the Buenos Aires Ranch. This contradictory situation was made possible by a far-flung corporate structure with investments in commercial real estate; other ranches; and a variety of agricultural, commercial, and industrial enterprises. Losses and expenses at the Buenos Aires Ranch served to write down gains in other branches of the company for federal tax purposes. In short, the ranch became a curious sort of business—a capital sink, so to speak—doing for a suite of subsidiaries and investors what it had previously done for the Dobson and Norton family businesses.

Nested within this larger framework was another set of motivations, however, which determined the ambition to restore the Buenos Aires to a grassland. After all, restoration was not necessary to the tax shelter. Large-scale revegetation efforts responded instead to an internal rent gap and, in so doing, appealed to the symbolic values of ranching. The forage productivity of the ranch had diminished some 75 percent since 1915 and perhaps as much as 95 percent since Aguirre's time. If one took "original" conditions as a fixed measure of potential carrying capacity—a premise that reached back to Jared Smith—then actual productivity was disturbingly low. Investing in restoration might ameliorate this, much like renovating a run-down building can close the rent gap for an urban landlord. In view of the declining profitability of cattle ranching and the rising opportunity costs associated with suburbanization (the larger rent gap described in chapter 5), this line of reasoning captured symbolic returns as well: Restoration was a high-stakes gamble to "save ranching."

Restoring the masked bobwhite was an ambition fraught with contradictions of its own. Having gone to the trouble of capturing birds, relocating them to Patuxent, and perfecting techniques of producing surplus offspring in captivity, the FWS could not give up without attempting to release them somewhere in the United States. Regional Director Michael Spear explained this to the national director in Washington, D.C., in a memo dated 3 September 1982: "The Service has much invested in the masked bobwhite, including field efforts in Arizona and Mexico and a major rearing program at Patuxent. . . . The Patuxent program is valid only as long as there are suitable habitats available into which the bird may be reestablished." In Patuxent the masked bobwhite was a *product* of human-controlled, industrial-style propagation. In Arizona it became an *agent:* Success required that the birds survive and reproduce on their own as they had prior to the cattle boom. If successful, the tremendous labor and expense involved in restoration would be erased by a social perception that nature was back in control. Indeed, human determinations *had to be erased* in order for the FWS to capitalize on the masked bobwhite's symbolic value as an icon of nature. Hoarded in cages in Maryland, the birds were like money stuffed in a mattress: they had to be thrown into nature to complete the process of valorization immanent to their form as a species of capital.

The combined result of these two restoration programs was dramatic but equivocal. The Victorio Company succeeded in restoring a grassland aspect on some sixty thousand acres, but only with the help of a prolific non-native perennial bunchgrass, Lehmann lovegrass, which still dominates large areas of the Buenos Aires Refuge today. Success was short lived, moreover: Mesquites resprouted or recolonized in some areas so rapidly that retreatment was necessary toward the end of Victorio's tenure. Likewise, the biologists succeeded in establishing a handful of small populations of masked bobwhites but only for two years, from 1977 to 1979. The birds declined after that, and by 1984 none could be found on the ranch.

The way these results were interpreted, first by the FWS and later by the public at large, was strikingly asymmetrical. The successful releases of 1977–79 all occurred in areas that had been cleared and seeded in the recent past. Researchers acknowledged that conditions in the Altar Valley had long been "totally unsuitable" for reintroduction (Brown and Ellis 1977: 7) and that mesquite removal had improved the habitat for masked bobwhites in important ways. Yet in their analysis, the role of the Victorio Company's activities was ignored, erased from consideration by a tacit judgment that habitat suitability was an ahistorical quality of the place. It was as though neither human acts nor contingent factors had played any role. When the birds perished after 1979, however, presumption quickly reversed: The failure of the summer rains was ignored, whereas Victorio's cattle were fingered as the cause of the decline. It was on this basis that the FWS and a coalition of environmental groups prevailed in establishing the refuge in 1985.

RESTORING GRASS

Wayne Pruett grew up on a large pure-breed and commercial Hereford ranch near Raton, New Mexico, where his father was manager. After earning a degree in agricultural economics at New Mexico State University, he moved to Phoenix in 1965 and took a job with a real estate firm that brokered ranch and farm loans for the Connecticut Mutual Life Insurance Company. Peter Wray worked in the same building for another firm, brokering investments in ranching for clients all over the country. Wray was married to an eastern heiress whose family connections helped in securing outside investors. In 1968 they formed the partnership that later became the Victorio Company. Pruett served as president and made his home at the Buenos Aires Ranch after 1972.

The agricultural side of the Victorio Company was divided into a cattle division and a real estate division. The real estate division owned the ranches and leased them at agricultural value to the cattle division, thereby insulating the latter from the costs of appreciating land values. Even so, according to Pruett, the cattle division was usually a break-even or money-losing component of the company, whereas the real estate division "never lost money."

The company's agricultural real-estate strategy had two parts. The short-term, "gain-on-sale" component sought "to acquire properties that are available on a distress basis . . . [and] to utilize the properties during the holding period, with the ultimate objective of selling the property at a profit" in a relatively short time.[1] Most of Victorio's ranch acquisitions were of this nature. The Buenos Aires was purchased under the second part of the strategy, as a "long-term holding . . . purchased in connection with The Victorio Company's cattle production activities." The huge Gray Ranch in far southwestern New Mexico, acquired in 1974, was another such Victorio property. At its peak, the Victorio Company ran twenty thousand mother cows and one hundred thousand stocker cattle on ranches in seven western states, plus a thirty-two-thousand–head feedlot.

All of Victorio's real estate investments were influenced by the tax policies described in chapter 5. Some of the capital generated by other divisions of the firm or raised from outside investors was converted into real estate to be sheltered from taxes. Pruett was a licensed appraiser, and he frequently did the appraisals on which accounting figures were based. In some cases, he reports, the depreciable improvements on a ranch appraised for more than the total purchase price, although the IRS would not permit *that* favorable a depreciation allowance. Once the improvements had been written off, the property could be sold and the proceeds reinvested in a like property with all capital gains tax deferred. Any annual losses reduced the tax liability of other, profitable divisions. With an upper marginal income tax rate of 50 percent, losses in the cattle division were, perhaps, half intentional: A dollar lost was, after taxes, only fifty cents lost, and gains through real estate appreciation might well cancel that out in the goodness of time.

Interwoven with fiscal advantages were the symbolic rewards associated with making the Buenos Aires into a model modern ranch: highly capitalized, intensively managed, and pitched to the highest echelons of American cattle production. Swollen by the flood of outside capital, the upper crust of ranching society cultivated its distinction in part through the breeding and exchange of "pure" or "blooded" stock. High cost and high status reinforced each other. Top registered bulls, like thoroughbred stallions, were known by proper name and could command astronomical prices. As in nineteenth-century England, high prices "bore little relation to the animals' practical utility. Their magnitude symbolized the significance of the enterprise as a whole, as well as the financial power of those who could afford to participate. . . . Possession of such animals required no pragmatic explanation. They were intrinsically desirable objects of conspicuous consumption" (Ritvo 1987: 55).

The Victorio Company went even further than most pure breeders by specializing in two breeds simultaneously: Hereford and Brangus. In common with commercial ranchers, Victorio bred for maximum weight gain, quality of meat, and other attributes defined by the market for slaughter animals (see chapter 4). The market for registered cattle, however, was commercial cow-calf ranchers, who sought purebred animals to improve the genetics of their herds. Victorio thus faced the additional challenge of maximizing prepotency, the genetic capacity to pass desired traits on to

Figure 11. One of the Victorio Company's Herefords on the Buenos Aires Ranch. Note the absence of mesquite trees, which the company had removed. (Courtesy of Wayne Pruett)

offspring. Even more than quality meat, prepotency cannot be directly observed in the living animal. A breeder's reputation contributes greatly to market value; the buyer pays for the prestige of owning stock of a particular pedigree. Victorio's Hereford operation excelled in this respect, winning the prestigious Carload Competition at the Denver Cattle Show four years in a row (1978–81; fig. 11).

Success in pure-breed ranching required exacting control over the mating of bulls to cows. From a technical standpoint, the challenge was similar to the one faced by biologists at Patuxent, except that range cattle are more difficult to control than caged quail. Victorio used a computer program to keep track of bloodlines and to determine which cows to expose to which bulls. On the ground, fences were necessary to confine individual bulls with selected cows; otherwise one could never confidently determine paternity. At purchase in 1972, the Buenos Aires Ranch consisted of sixteen pastures, as it had in 1959.[2] For two years, Victorio employed a full-time crew solely for fence construction. Total fencing on the ranch rose to 370 miles, partitioning off some seventy separate pastures. Because every pasture had to have at least one water source, the Gills' waterline from Figueroa to Secundino was revived, as were the Bailey Wash headgate and dike and the complex of pumps and dams at Aguirre Lake. Two full-time bulldozer operators worked to clean and improve stock tanks. Eventually, a hundred stock tanks and twenty-one wells serviced the range of the Buenos Aires.

The high value of the livestock and the emphasis on breeding made good forage imperative, both to keep cows in prime condition for breeding and to maximize the growth of calves. The impetus to tackle the mesquite problem must be understood in this light. "You've got to be a grass farmer first and a cattle breeder second," as Pruett liked to say (*Arizona Daily Star,* 23 September 1979). The program he led was, indeed, a kind of brute farming, a belated realization of the Division of Agrostology's vision of a cultivated range. The technology was less precise and the outcome less controllable than in crop agriculture, but the force exerted was that much greater. In the 1960s, workers on the neighboring Santa Margarita Ranch had chained twenty thousand acres of its range by pulling a sea anchor chain between two bulldozers to level the brush and disturb the soil. At the time, diesel fuel cost seventeen cents per gallon and calves were bringing eighty cents per pound, according to the ranch manager's recollections. At $7.50 per acre, the costs were paid back, with interest, in the form of increased grazing capacity. After the oil crisis of the early 1970s, these ratios no longer held, but Victorio, like the Gills before, applied the neighbor's practice on a much grander scale anyway.

An all-out assault on mesquite trees was launched in 1973 (fig. 12). Victorio benefited from one lesson from the earlier efforts: Chaining had not killed many mesquites but had simply pushed them over or broken them off at ground level, leaving the root crowns to resprout. Victorio used a "stinger" attachment on a bulldozer to probe underground and cut the taproot, a practice known as grubbing. Grubbing was considerably more expensive per acre than chaining (and chaining was also done), but it killed the trees and left behind small depressions to catch runoff and thereby help slow erosion. Dozens of these "potholes" can be seen in some areas to this day. Larger trees were sold for firewood; smaller ones were pushed into large piles or windrows to be burned or left as wildlife habi-

Figure 12. Using a sea anchor chain, stretched between bulldozers, to remove mesquite trees on the Buenos Aires Ranch. (Courtesy of Phil Ogden)

tat. Compacted soils were then ripped on contour to increase water infiltration and soil aeration.

After a pasture had been treated, grass seed was sown, usually from an airplane. The seed mixtures contained both native and non-native species; both cool season and warm season species were used in hopes of establishing good forage throughout the year.[3] Johnson grass and blue panic grass *(Panicum antidotale)* dominated the seed mixture used in bottomland areas, whereas blue panic, green sprangletop *(Leptichloa dubia)*, and three lovegrasses—plains *(Eragrostis intermedia)*, Boer (*E. curvula* [Schrad.] Nees var. *conferta*), and Lehmann—made up the mix for upland pastures. Treated pastures were rested from grazing for at least one and typically two growing seasons to allow for seedling establishment.

Determining the exact acreage that was treated in this way is surprisingly difficult. Several documents exist, but the reliability and extent of overlap among them is impossible to determine. The best estimate is that sixty thousand acres of mesquites were chained or grubbed,[4] about equally divided between the north and south ends of the ranch; roughly two-thirds of this work took place on State Trust lands, with the remainder on deeded land. Generally work progressed from north to south on the ranch; one source recalled that the ratio of non-native grasses (Lehmann and Boer lovegrass) was higher in the earlier years. This may explain the higher prevalence of these species on the north end of the refuge today. Almost all of the ranch's private land was treated, reflecting two advantages from the point of view of the Victorio Company. First, the deeded land was concentrated in the loamy bottomlands, where the most dramatic increases in forage production could be achieved (especially in the form of Johnson grass). Second, no permits were required to work on deeded land.

Despite these measures, mesquites persisted in sprouting across much of the Buenos Aires Ranch, with the exception of some bottomland areas where Johnson grass grew so thick that mesquite seedlings could not survive to maturity. On a much smaller scale, Victorio experimented with other techniques in hopes of finding a cheaper and more effective means of control. In 1977 the ranch entered into an agreement for experimental spraying of 2,4,5-T on four thousand acres of State Trust land. Researchers from the University of Arizona, USDA, Dow Chemical, and Eli Lilly oversaw the experiment and provided the chemicals at no charge to Victorio. The project became snarled in public controversy, however, and it never yielded the sought-after "100 percent kill." (After Tucson environmentalists criticized the mesquite removal program, Victorio made a policy not to remove trees adjacent to the highway to minimize public exposure to their efforts.) Fire was also used to help remove snakeweed and burroweed from two small areas in 1979–80, with a beneficial impact on grass production. Burning could only occur in areas already treated, however, because a stand of grass was needed to carry the flames.

As the Gills had done before, Victorio also turned its bulldozers to the task of erosion control. After chaining or grubbing a pasture, operators performed "dirt work," pushing up berms of earth across incipient channels or anywhere erosion was evident. Literally thousands of little impoundments were created in this way to catch water and sediment after a rain. On a larger scale, spreader dams were reconstructed across Puerto-cito Wash and other tributaries of the Altar Wash to draw runoff onto the flats, where it would irrigate the Johnson grass. Wayne Pruett boasts that no water flowed off the north boundary of the Buenos Aires Ranch when this system was in place. The spreader dams did wash out, however, during the huge floods of 1983, and they were never repaired.

The final component of Victorio's management was rest-rotation grazing, facilitated by the new fences and water sources. Each pasture received a period of grazing, followed by a longer period of rest; typically, a herd rotated through a series of pastures on a one- or two-year sched-ule. Unlike earlier management practices, rest-rotation could accommo-date interannual variations in rainfall and forage growth. The degree of grazing in a pasture could be monitored and the herd moved when a certain limit was reached. Rest-rotation is now standard practice on many southwestern ranches, but in 1972 it was still novel.

Victorio's range-improvement program borrowed many tools from the Gills' earlier efforts, but the aggregate effect was in key respects the opposite. The Gills' grazing, combined with the drought of the 1950s, had diminished grass cover, encouraged the spread of mesquites, and increased runoff rates. Victorio found that complete removal of mesquites and resto-ration of grass cover would render downstream stock tanks inoperable, "restoring" the hydrological balance of 1880. There was simply not enough water for both the tanks and the grass, and the grass got it first. (The same correlation was reported by a former manager of the Santa Margarita Ranch and confirmed by subsequent events on the Buenos Aires Refuge.) Although Victorio calculated that one cow could be added to the range for every fifty acres of mesquite that were cleared, they never came close to the peak stocking rates of the earlier period. The estimated carrying capacity at the end of their tenure was twenty-five hundred to twenty-seven hundred cattle, or not more than fifteen head per section.

RESTORING QUAIL
REINTRODUCTION EFFORTS IN THE 1970S

After an absence of some eighty years, the masked bobwhite returned to the Buenos Aires in the midst of Victorio's range-improvement program and, indeed, because of it. The two restoration efforts did not inform each other, however, because of institutional, professional, and social barriers separating biologists and wildlife agencies from ranchers and range scientists. The symbolic opposition between cattle and the masked bobwhite reinforced these lines of division.

Reintroducing the masked bobwhite to the Altar Valley was, and remains, a matter of hitting a moving target: Both the birds and the habitat are dynamic, and fitting them together is an elusive challenge. Bred in captivity, the masked bobwhites' "natural" instincts are always in question. Their viability for release may diminish because of incipient domestication as well as inbreeding, poor health, or conditioning. Accordingly, one aspect of reintroduction work in the 1970s consisted of "efforts to upgrade the stock produced in captivity" (Ellis et al. 1977: 345) through modified conditioning and release protocols. The question of habitat was equally troublesome. Not only were habitat requirements poorly understood, but actual conditions varied over time in response to rainfall, temperature, grazing pressure, and range manipulation. Selecting and studying release sites thus made up the second major aspect of reintroduction research. When released birds perished without reproducing, it was inferred that one or the other aspect of the program was to blame. Conversely, when the releases succeeded after 1976, researchers concluded that they had found the "natural" combination of habitat and birds.

Early FWS reintroduction efforts were no more successful than their predecessors. From 1970 to 1974, nearly two thousand masked bobwhites were released at sites in the Altar Valley, mostly located in the mesa lands of the Santa Margarita Ranch, Las Delicias Ranch, Rancho Seco, and on the USFS allotments of the Buenos Aires Ranch, under agreements that excluded grazing. Chicks were shipped by air from Patuxent at two weeks of age (figs. 13 and 14). Some were released immediately; others were held for up to three months. Later, additional acclimation time in pre-release pens was provided. "[O]ne case of successful reproduction was documented, and a few birds were known to have survived for up to one year," but results otherwise were "discouraging" (Johnson and Hoffman n.d.: 4; cf. Brown and Ellis 1977: 10; Tomlinson 1972a). Most birds could not be found two months after release. Predation, especially by raptors, was high; malformed beaks and "excessive plumage wear" were deemed

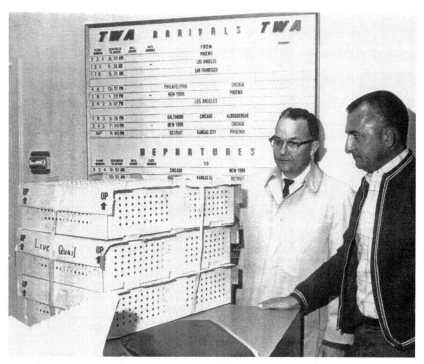

Figure 13. Crates of masked bobwhites arrive at Tucson International Airport from the Patuxent breeding facility, 2 March 1970. Ray Erickson (left) was director at Patuxent; Roy Tomlinson (right) conducted most of the early federal field research on masked bobwhites in Sonora. (Courtesy of the U.S. Bureau of Sport Fisheries and Wildlife)

contributing factors (Ellis and Serafin 1977: 27). After some of the birds migrated toward lower ground, researchers concluded that bottomlands were their preferred habitat (Brown 1989: 135–136).

The failed experiments of 1970–74 almost killed the masked bobwhite program. According to several of the researchers, who were interviewed in 1998, supervisors in the FWS regional office in Albuquerque were pessimistic and had begun to view the program as a liability. Clearly there were tensions between field researchers and office administrators. David Brown, an AGFD biologist and a member of the Recovery Team at the time, recalls that the wild birds captured in 1970 were all shipped to Patuxent over his objections. Brown believed that releasing at least some of the wild birds was necessary to advance the research, and the decision

Figure 14. David Brown (left) and Roy Tomlinson observing masked bobwhites in pens prior to release, 3 March 1970. The release site was located on the Santa Margarita Ranch, where mesquite trees had been bulldozed in the 1960s. (Courtesy of the U.S. Bureau of Sport Fisheries and Wildlife)

to use all of them for breeding instead indicated to him that "justifying Patuxent's existence" had become the overriding focus of the program.[5]

In late 1974 the FWS hired David Ellis, who had just completed a doctoral thesis on behavioral development in eagles, to take charge of masked bobwhite work in the field. He recalls learning quickly that the project was in jeopardy and resolving that he would give it two years' work. He rapidly got Patuxent's attention: Many of the birds they were producing, he reported, were too disabled from improper debeaking and feather damage to be of any use in releases. He announced that he would kill them rather than allow them to attract predators to the other, better conditioned birds.

For adult birds, Ellis and his team developed a twenty-four-day "training" program consisting of gradual exposure to the outdoors; harassment by dogs, humans, and a trained hawk; and a call-box system to instill covey behavior. "During training, the birds rapidly improved in general mobility, in coordination within the covey, and in ability to hide and avoid predators. Dogs proved very useful in simulating mammalian predation, and the quail quickly learned when to hold and when to flush. . . . [T]he hawk was very useful in evaluating whether birds were strong enough for release." Quantitative standards were then imposed to cull unfit birds: "To be acceptable for final release, each bird had to: (1) have at least 90 percent of its wing surface intact, (2) have at least 50 percent of its tail surface present . . . (3) have no serious deformities or injuries, (4) be able to fly strongly at least 200 m[eters] on the second flush, and (5) be able to use cover effectively and avoid predators" (Ellis et al. 1977: 349). Approximately 70 percent of the birds met these criteria. For chicks, experiments were conducted with a variety of "surrogate" parents, of which male Texas bobwhites *(Colinus virginianus texanus),* captured from the wild and vasectomized to prevent interbreeding, were found to be most successful.[6] All of the masked bobwhites released on the Buenos Aires Ranch in 1976 and 1977 underwent one or the other of these conditioning programs (Brown and Ellis 1977: 11).

Concurrent with the conditioning experiments, researchers were refining their judgments of habitat needs. Agreements were made with Pruett and Wray to release birds into bottomlands along Puertocito Wash that were being rested following recent manipulations. Gently sloping hills, also recently cleared and seeded, rose on either side. Before manipulation, these areas were dominated by mesquites that were quite large in the bottomlands and smaller on the uplands; this likely explains why they had not been selected for releases earlier. Following manipulation, they presented an entirely different aspect: thick stands of giant sacaton and Johnson grass in the bottomlands, young green grasses and forbs in the uplands, and brush piles and remaining trees interspersed throughout. Tomlinson (1972a: 300) had observed that the resulting "edge effect" appeared to attract masked bobwhites in Sonora, providing protective cover in the bottomlands and abundant food in the more open grasslands above.

Released into these pastures, the call-box-conditioned and foster-parented birds finally succeeded in reproducing. With a tone of vindication, the researchers wrote:

> Many of the birds released in 1976 survived into the winter. With the onset of the 1977 summer rains a population estimated at around 30 birds was located near the 1976 release sites in Altar Valley. On October 4, 1977, a pair of masked bobwhites was observed with at least three chicks. These observations constitute (1) the first demonstration of significant over winter survival, (2) the first establishment of a "wild" breeding population, and (3) the first observations of progeny reared in the wild by fully independent stock of propagated origin. These events substantiate the feasibility of re-establishing the masked bobwhite. (Brown and Ellis 1977: 10–11)

By 1979 five small populations had been established on the Buenos Aires Ranch, and the number of observed calling males had increased to seventy-four from twenty-one in 1977.

The successes prompted rapid formulation of plans for a refuge, with the Buenos Aires Ranch as the obvious first choice. Less than three months after the 1977 sightings, the Masked Bobwhite Recovery Team completed the first official Recovery Plan for the subspecies. Restoring the masked bobwhite, it began, was "perhaps readily feasible, perhaps impossible" (Brown and Ellis 1977: vii). The plan unequivocally attributed the extirpation of the masked bobwhite to livestock grazing (ibid.:

6) and, after describing the successful releases, concluded, "It is as yet uncertain that these coveys will survive indefinitely. Although sizeable areas of the habitat type preferred by the conditioned birds are presently managed by a progressive and enlightened livestock operator, these potentially essential habitats are not managed by an agency with wildlife management responsibilities. Even if these coveys result in the establishment of a self-sustaining population, that population will remain in jeopardy because of the potential for excessive grazing" (ibid.: 12).

The plan conceded that a habitat reserve in Mexico was "more biologically feasible" than reintroduction in the United States, but it asserted that "severe social, economic and political problems" there might someday "negate any previous commitment to a program to benefit an economically minor subspecies of wildlife" (ibid.). Therefore, "the acquisition of a habitat reserve and the establishment of a population in Arizona offers the better opportunity to prevent the bird's extinction" (ibid.). Signed by FWS Director Lynn Greenwalt in early 1978, the Recovery Plan signaled that the bureaucracy again believed in the masked bobwhite. According to researchers and former ranch personnel, acquisition of at least a portion of the Buenos Aires Ranch was an FWS objective from about this time forward. Approximately five thousand acres was viewed as the amount of land needed.

A drought at the end of the 1970s threw a shadow over the earlier successes, however. Measured against monthly averages, the rains failed in September 1979, July 1980, and August 1981 (Karl and Knight 1985). According to researchers' field notes, less than five inches of rain fell at the Buenos Aires headquarters in the summer of 1979, and less than two inches farther north at Pozo Nuevo. Summer drought has a strong, lagged impact on masked bobwhites. "The amount of summer precipitation determines the quantity of herbaceous vegetation produced and correlates with reproductive success. However, . . . total quail numbers are probably determined more by overwinter carryover than nesting success. This variation in survival is dependent on the amount of residual ground cover, so that population size is related to grazing intensity and the amount of rain received the *previous* summer" (Brown 1989: 138, emphasis in original). Releases were discontinued after 1979, and surveys in 1980 and 1981 found only fifty-one and twenty calling males, respectively (Goodwin

1982: 37).[7] By 1982 only five calling males could be located (Ough and deVos 1982), and in a 1984 survey no birds were seen or heard (Mills 1984). In 1985, shortly after the FWS took over the Buenos Aires, biologist Steve Dobrott did see a single masked bobwhite, the only known descendent of the 1970s' releases.

INTERPRETING THE CRASH
THE SCIENCE AND SYMBOLISM OF CATTLE AND QUAIL

In 1904 Herbert Brown had attributed the masked bobwhite's demise in Arizona to the *combined* effects of cattle grazing and drought. After decades of failed attempts at reintroduction, ornithologists in the 1960s chose to blame grazing and downplay (or simply not mention) drought, a shift in emphasis that reflected two judgments. First, drought came to be viewed as a natural phenomenon in the desert. Presumably the bird had survived dry periods before 1890, so in theory lack of rainfall could not be the cause of its extirpation. Second, excluding drought focused attention on the continued degradation of wildlife habitat by cattle grazing, in keeping with broader shifts in environmental perceptions at the time. The two positions were mutually reinforcing: The more one viewed drought as "natural," the more cattle appeared "unnatural."

Scientifically, however, both drought and grazing were necessary to explain the masked bobwhite's disappearance because both had clearly been involved in the 1890s, and no controlled experiments had shed new light on the subject. It was simply not possible to know whether the masked bobwhite might have persisted in the absence of one factor or the other. The typical way of expressing this in the research literature of the 1970s was to write that "[c]ircumstantial evidence associating the demise of masked bobwhites with heavy livestock grazing is overwhelming" (Brown and Ellis 1977: 5). Indeed, it is. But what allowed the birds to persist in Sonora, where grazing and drought were (and are) likewise present? How much grazing and how much drought could the birds tolerate? There were (and still are) too many variables to answer these questions conclusively. The history of the Buenos Aires amply documents that cattle have vastly greater impacts on vegetation during drought than under "normal" conditions. Some rainfall threshold might be hypothesized, below which cattle and masked bobwhite competed directly, and cattle, no matter how miserable, prevailed. This threshold, however, could not be specified with any precision, nor could the release experiments of the 1970s answer the drought-vs.-grazing question unequivocally. Fences meant nothing to the bobwhite; the birds frequently moved in and out of pastures being grazed.[8] Under the lease arrangements, the Victorio Company had the option to graze release pastures in times of drought, although interviews with researchers and former ranch personnel yielded contradictory accounts of whether, when, and how much this occurred.[9]

The only documented evidence on this crucial point—indeed, the only real analysis of the Buenos Aires release experiments—is found in an internal contract report written by John Goodwin in 1982. It indicates that grazing did occur in release areas *both before and after 1979,* and the data it contains are inconsistent as to the effects of grazing on the bobwhite. There were five populations spread out across the ranch. All occupied areas manipulated before or during the releases, and all diminished in numbers after 1979. Otherwise, the findings were mixed. The largest population developed in a pasture experiencing light grazing in 1978–79; after heavy grazing and a fire on part of the pasture in the summer of 1980, the population declined. A second group increased slightly from 1978 to 1979, despite heavy grazing, then declined in the absence of grazing. A third declined, then rebounded under light grazing from 1977 to 1979, then tapered off after a fire and heavy grazing in 1980 (Goodwin 1982: 13, 37). The other two populations were much smaller and were not censused as carefully (Goodwin 1982: 17); for one the grazing history is unknown, whereas the other crashed following heavy grazing in the winter of 1979. The only direct evidence of cattle impacts on quail offered in the report was the observation that one group of birds relocated away from a heavily grazed area between 1980 and 1981 (Goodwin 1982: 33).

The conclusion drawn by the researchers was a negative one. The habitat on the Buenos Aires Ranch, although better than any other in the United States, was still "marginal" for the quail. It followed that *any* grazing pressure posed a threat to the birds, especially in periods of drought.[10] Removing cattle was therefore a necessary *but not sufficient* condition for ensuring their survival. This conclusion, however, remained basically circumstantial in its evidence. It would have been more accurate to say that no further scientific conclusions could be reached one way or the other without gaining strict control over grazing so that its impacts could be carefully analyzed.

Five other factors confounded the research results even further, although only the first received any consideration in Goodwin's report or in the 1977 Recovery Plan:

First, all of the release pastures had been intensively manipulated—chained, grubbed, seeded, or burned—shortly before or during the experiments.[11] Habitat manipulation is known to benefit quail by producing a flush of new growth, especially of annual grasses and forbs. The effects are transient, however. (Manipulations had also occurred in the area on Rancho El Carrizo where wild masked bobwhites were found in the early 1960s [Tomlinson 1972a: 297], which explains why the habitat there "stood out like a fire hydrant in a parking lot," in Jim Levy's words.) Similarly, Goodwin (1982: 25) credited the brush piles left by the Victorio Company with providing important cover for masked bobwhites during the 1977–79 releases, although he lamented that the larger mesquites had been sold for firewood and that termites would rapidly decompose what was left. He warned that "a consistent burning program will eventually remove all brushpiles and young shrubs necessary to support a bobwhite population" (1982: 28).

Second, according to Stoddard (1931), two of the bobwhite quail's preferred foods are Johnson grass seed and panic grass seed, both of which the Victorio Company sowed on release pastures by the hundreds of pounds. The effects of this seeding on masked bobwhite survival were unknown.

Third, the value of call-count data was questionable because new birds were released each year. How many were "wild" offspring and how many were captive bred was unknown or at least unreported.

Fourth, the number and age-class of released birds also varied between 1975 and 1979; declines after 1979 might have been attributable to differential survival rates between chicks and adults.[12] Only chicks were released in 1979; adults or juveniles might have survived better.

Fifth, by increasing soil moisture in the bottomlands, Victorio's spreader dams may have strongly influenced habitat conditions during the experiments (cf. King 1998). Recall that breeding behavior is triggered by microclimatic humidity levels (see chapter 1).

In short, the researchers could not assess the relative importance of management decisions (livestock grazing and range manipulations) and natural factors (primarily rainfall) in determining the outcome of their experiments. They had proved that reintroduction was possible on the Buenos Aires Ranch, but little else. The end of the experiments forced this scientific uncertainty to a bureaucratic crossroads. How was the crash to be explained? If it was the result of drought, then the earlier successes were empty and masked bobwhite recovery apparently impossible. If grazing were the cause, however, then a powerful argument for the creation of a refuge could be made. At this point, the symbolic opposition between cattle and quail superseded the scientific problem of multivariate causality.

Two events in 1983 strongly influenced the future of the Buenos Aires Refuge, though in somewhat opposing ways. First, Pruett and Wray's Victorio Company, reeling from a recession and overextended on several commercial real-estate ventures, transferred the Buenos Aires Ranch (along with the even larger Gray Ranch) to Mexican billionaire Pablo Brenner's American Breco Company in settlement of a debt. Looking to recover his losses, Brenner stocked the ranch with steers and announced his intention to sell. Within the FWS, this heightened the urgency of acquiring the Buenos Aires Ranch. Any new owner, including Brenner, might refuse to cooperate with further research or graze the ranch more aggressively and diminish its quality as bobwhite habitat. According to a 1985 memo from Goodwin to the FWS regional director, Brenner's steers badly overgrazed prime habitat near headquarters.

Second, a tremendous flood struck in the fall of 1983, wiping out the spreader dams, headgates, and dikes that the Victorio Company had resuscitated from the days of Roy Gill. Brenner did not restore these developments, and neither has refuge management. In retrospect, it appears that the suitability of the Buenos Aires for masked bobwhites may have diminished immediately before the creation of the refuge. Biological issues, however, scarcely arose during the political process of acquiring the Buenos Aires.

At first, the FWS sought to buy only a small portion of the Buenos Aires Ranch—the bottomlands south of Arivaca Road—in keeping with the Recovery Plan's call for approximately five thousand acres. Brenner, however, refused to break the ranch into pieces. It was all or nothing, he announced, and his price was $9 million. Cost was not a significant obstacle for the FWS bureaucracy because the money for the purchase would come from a large federal fund earmarked for conservation and replenished by offshore-drilling royalties. All that was needed was congressional approval and an appraisal to meet the legal requirement that only fair market value be paid. Congressman Morris Udall was instrumental in securing a $5 million authorization in the fiscal year 1984 budget and an additional $2.5 million the following year. Two independent appraisals conducted in the spring of 1984 assigned values of $6.8 million and $7.1 million, respectively. A third appraisal, conducted internally by an FWS realty official, arrived at a value of $8.9 million. Brenner and the FWS came to terms around this figure, and by the end of 1984 the deal lacked only the signature of the U.S. Secretary of the Interior, William Clark. Calling the cost of the proposed refuge "preposterous," Clark stalled the final approval.

The political campaign for acquisition relied on a simplified presentation of the biological issues surrounding the masked bobwhite. The successful reintroductions of 1977–79 were advanced as proof that release techniques had been perfected and that the masked bobwhite could survive on the Buenos Aires. In a letter to Senator Dennis DeConcini in July 1984, Acting Regional Director Ellis Klett wrote, "Our field efforts eventually resulted in the successful reestablishment of the masked bobwhite in Arizona." (One month later a survey would find no masked bobwhites on the Buenos Aires [Mills 1984].) The location of the ranch within the bird's historical range was strongly invoked. The crash of released populations after 1979 was presented as evidence of the need to create a refuge and remove cattle. "We know the birds can survive, but not until we get control over the management of the pastures on the ranch," one FWS official told the *Arizona Daily Star* (2 February 1984). The ambiguity of the scientific research—the fact that no one really knew what the bobwhite's prospects were on the Buenos Aires—was elided, erased by the symbolically powerful and politically palatable notion that the bob-white would survive "naturally" if the "unnatural" element—cattle—was removed. The National Audubon Society, two Arizona Audubon chapters, the Nature Conservancy, and the Arizona Wildlife Federation joined the AGFD, Governor Bruce Babbitt, and Arizona's congressional delegation in supporting acquisition.

Proponents routinely cast the refuge proposal in the larger context of endangered species restoration as a whole. The Endangered Species Act mandated the protection and restoration of many species, but in most cases this was (and is) profoundly difficult. Most wildlife species are difficult or impossible to breed in captivity; others are too costly, in political or economic terms, to make restoration feasible. At a symposium convened at the University of Arizona in December 1984, Goodwin summarized the situation:

> The only factor I can see against the Buenos Aires is the expense. Yes, it is expensive. But, the masked bobwhite is one of the very few endangered species we may actually be able to save. And by law the U.S. Fish and Wildlife Service must pay only fair market value. We are going to lose a lot of other species within the very near future—most of them simply because we do not have the time, research results, money, and manpower to save them. The background work has been done on the masked bobwhite, and we are in a position to save it. (Stromberg et al. 1986: 9; cf. Ellis et al. 1977)

Here was a species that apparently could be brought back from the brink of extinction, and the acquisition of a single property would be sufficient.[13] It might even compensate, symbolically at least, for the many species whose prospects were seen as dim. Success could never be guaranteed, but chances seemed better than in most other cases. As the first Recovery Plan put it, a refuge for the masked bobwhite would uphold "the *spirit* of the Endangered Species Act" (Brown and Ellis 1977: vii, emphasis added), regardless of whether the bird actually survived.

Opponents of the proposed refuge were almost all local residents who were concerned about the economic and fiscal implications of having the Buenos Aires removed from private production. They focused their attention on the high price tag to taxpayers during a period of deficit spending and on potential impacts on the local tax base and school district. A few speculated publicly that the federal government had ulterior motives for wanting the refuge: as a getaway for Washington bureaucrats, for instance (*Arizona Daily Star,* 22 February 1985).

Only one person openly questioned the biological claims advanced by the FWS. Professor Stephen Russell of the University of Arizona, an expert on southwestern birds, pointed out at public hearings that the Buenos Aires was on the margin of the masked bobwhite's historical range and might therefore never have been perennially suited to the bird (FWS 1984a; cf. Stromberg et al. 1986: 18–21). "[I]t is very likely that an apparently well established and managed masked bobwhite population at Buenos Aires Ranch could disappear with unusual cold, drought, disease or whatever. . . . I don't think a self-sustaining masked bobwhite population can be maintained on the Buenos Aires Ranch without drastic management activities" (FWS 1984a: 36, 38). Russell also noted several vague or misleading claims made by the FWS, and he argued that the money would be better spent to conserve wild masked bobwhites in Mexico. Russell's points have been borne out by subsequent events (see chapters 7 and 8), but they did little to slow the move to create a refuge. As several local residents observed at the hearings, the real decisions had already been made, and gathering public comment was merely a formality (FWS 1984a, 1984b). After William Clark resigned as the U.S. Secretary of the Interior in February 1985, his successor, Donald Hodel, quickly approved the acquisition of the Buenos Aires Ranch.

CONCLUSION

The Victorio Company's investments in the Buenos Aires Ranch brought about a remarkable change from the conditions of the 1950s and 1960s. More than ninety square miles of mesquite-dominated scrubland were restored to grassland, with far-reaching effects on runoff, vegetation, and wildlife habitat. Greater grass cover, erosion control measures, and spreader dams together approximated the pre-entrenchment hydrology of the valley, at least temporarily. It is fair to say that without this work, the masked bobwhite reintroductions of the late 1970s would not have been successful. Restoration, however, was neither complete nor permanent. Victorio's program did not restore the composition or productivity of pre–cattle boom vegetation. Lehmann lovegrass came to dominate much of the treated area, sometimes in near monotypes. Although it competes well with mesquites and other brush and it creates the appearance of lush grassland, Lehmann lovegrass is viewed as an "invasive species" that may adversely affect native grasses and certain wildlife. (The same criticisms can be made of Johnson grass, which has largely replaced the native giant sacaton in the bottomlands.) Moreover, even Lehmann lovegrass could not prevent mesquites from returning to the uplands in subsequent years. Finally, areas that were not treated by the Victorio Company were not restored; indeed, they retain today the mesquite and shrub plant communities that all parties agree are inferior for soil stability, livestock production, and wildlife.

Restoration of the masked bobwhite was likewise incomplete and transient. Official listing of the subspecies under the Endangered Species Act subsumed reintroduction efforts in a bureaucratic system of value production, backed by laws and funded by taxpayers. With increasing precision the FWS subjected the masked bobwhite's "nature" to scientific analysis, very much as cattle had been since the turn of the century, and to the same end: maximum production of highest-quality animals. After nearly four dozen failed reintroductions in New Mexico, Arizona, and Sonora, dating back to 1937, the successes of the late 1970s represented a major and long-awaited breakthrough. But it could hardly be presumed that replication would be automatic or easy.

The symbolic value of the masked bobwhite depended on its abstraction from human society and its construction as a "wild" and "natural" species. The instrumental knowledge and techniques developed by researchers were sanctified by the essentially ahistorical and transcendent object of their efforts. The bird's range, breeding habits, diet, and habitat needs were constructed as fixed, given, and "natural," even as they were systematically manipulated. All of the human interventions— captive breeding, conditioning and release techniques, range improvements, and erosion control—and all of the contingent or temporary habitat factors—rainfall, broadcast grass seed, altered soil moisture levels, and short-term vegetation responses to disturbance—were effaced, rendered invisible in the white light of timeless Nature. Suddenly the birds were the agents, reproducing "on their own," transferring their genetic-evolutionary uniqueness to the Buenos Aires as a one-of-a-kind place, "the only habitat in the United States presently believed to be capable of supporting a viable population" (FWS 1985). When the birds declined after 1979, a culprit was ready at hand. Cattle, also construed in a time-less fashion, were somehow "naturally" incompatible with the masked bobwhite, despite the fact that the sole remaining wild populations occupied—and still occupy—a working cattle ranch. The scientists work-ing for the FWS and the AGFD did not themselves subscribe to such a simplistic understanding, but in the political realm the subtleties of biol-ogy were quickly lost.

Science was not altogether innocent, however. After all, it was the taxonomic system from which the very idea of endangered species issued. If the Endangered Species Act has been successful in preserving and protecting wildlife, as most ecologists and environmentalists agree, it is not by virtue of its ecological coherence but because of its legal and political precision. It makes the living creatures themselves into a form of property that can be defined, counted, banded, propagated, litigated, and valued along the same lines as commodities; and commodities are something our legal and political system is well equipped to adjudicate. This is especially true when conjoined with the charismatic appeal that some wild animals hold in our culture, an appeal that was itself critical in passing the Endangered Species Act in the first place. Science, sentiment, and law combine to *fetishize* endangered species. "Listed" species become invested with a transcendent value that is understood to inhere in their *bodies* rather than in the ecosystems of which each is but a small part, just as the value of commodities appears to be a quality of the objects themselves rather than of the social relations that produce them.[14] The scale against which this value is measured, moreover, is global, even universal: They are not just rare in one particular place but in all the world, utterly irreplaceable. Thus, when citizens (mostly locals) argued that the masked bobwhite could not justify $9 million of public expense, supporters of the refuge (mostly nonlocals) responded that any species is priceless and that abandoning the bird on such grounds would set a dangerous precedent. The two sides argued right past each other: For one, the masked bobwhite was just another bird; for the other, it was akin to a work of art. For both sides, however, *the bird* was the issue because species—and in this case subspecies, long known as races—constituted the conceptual frame through which the state and the law apprehended nature. Had it not been for the masked bobwhite's symbolic opposition to cattle, moreover, the FWS would probably not have found so much support. The creation of the Buenos Aires Refuge was a surpassing symbolic victory for environmentalists who were agitated by the historical damage and political power of ranching in the West. In the following chapters we will see whether it was anything more than that.

producing a state of nature

In the early summer of 1985, Pablo Brenner's employees rounded up the Buenos Aires herd one last time. When the FWS took possession on 1 August, the cattle were gone. Ever since, the absence of livestock has dominated perceptions and opinions of the refuge. Critics, especially ranchers, view it as a waste: of forage that could be grazed, of expensive range improvements, and of taxpayers' money. When I told one area rancher that I was studying what the FWS has done with the Buenos Aires Refuge, he interrupted: "I'll tell you what they've done: nothing!" But "nothing" is something to refuge staff and supporters, who uphold the significance of the Buenos Aires Refuge as the largest ungrazed grassland in Arizona. Far from a waste, they see livestock exclusion as productive: of wildlife values, conservation values, and resource values (and they use these terms). Overlooking or unaware of the history of the ranch, they often pronounce the Buenos Aires Refuge a "natural" landscape, a remnant or re-creation of "original" desert grassland ecosystems elsewhere defiled by cattle. For them, the symbolic value of the masked bobwhite has transferred to the refuge as a place.

Although the exclusion of livestock has unquestionably affected the ecology of the Buenos Aires Refuge, it is almost impossible to evaluate the effects in detail. The FWS asserts that its management is restoring native grasses and suppressing invasive species such as mesquite and Lehmann lovegrass (FWS 2000), but there are no solid data to support these claims. Vegetation transects, monitored irregularly since 1986, indicate that refuge management has had no statistically significant effect on the composition or density of vegetation on the Buenos Aires Refuge (Geiger and McPherson 2000). Since the refuge was acquired, Aguirre Lake and about two-thirds of the ranch's earthen stock tanks have ceased to fill. How much this is due to livestock exclusion (which presumably has diminished runoff rates) and how much to lack of tank maintenance is unclear; probably both are involved. The disappearance of these surface waters may be seen as a partial return to the hydrological conditions of 1880, except that the Altar Wash and other arroyos have not healed, meaning that the alluvial water table in the bottomlands remains some fifteen feet below its pre–cattle boom level. The full ecological significance of this is unknown, but recent research suggests that it may decisively impact the masked bobwhite's prospects (King 1998). The refuge's primary management tool, prescribed fire, has been extensively applied, but no comprehensive baseline data were collected, so it is impossible to determine what changes, if any, have resulted. Areas that the Victorio Company did not clear and reseed, for instance, remain in mesquite-shrub condition, with inadequate grass cover to carry a fire. Refuge biologists have concentrated their work on the masked bobwhite and its habitat needs, and outside scientists have likewise tended to focus on individual species. To date, the refuge has produced surprisingly little published research about broader processes of ecological change. For all these reasons, ecological criteria cannot be used here as the primary tools for understanding the transformation from ranch to refuge.

In terms of relations to nature on the Buenos Aires, creation of the refuge has simply replaced cattle with quail. The FWS strives "to *optimize* production of masked bobwhite on [the] Buenos Aires refuge" (FWS 1986a: 2, emphasis in original), just as ranchers did with cattle. Captive propagation of chicks employs techniques derived from commercial poultry production; as with purebred cattle, the aim is to produce genetically "pure" animals while avoiding the negative side effects of inbreeding. After release, the birds are monitored and studied to discern their habitat needs, much as early range researchers sought to identify valuable forage species for livestock. Refuge staff work to manipulate and improve habitat for masked bobwhites, as previous owners did for cattle. Certain sociospatial hierarchies have also transferred directly to the refuge. The main ranch house has become a visitors' center, symbolically installing the public as the new owner. The ranch manager's house is now the refuge manager's, supervisors occupy previous foremen's houses, and volunteers make use of an old cowboys' trailer.

The principal discontinuity between ranch and refuge lies not in cattle or quail, nor in relations to nature, but in larger social structures of production: private markets and public agencies, respectively. The creation of the refuge entailed two changes in the circuits of capital accumulation affecting the Buenos Aires. The first was a change in the form of mediation through which value was produced: Bureaucratic capital replaced economic capital as the dominant form of social reproduction on the Buenos Aires. The second was a change within the process of bureaucratic capital accumulation itself. It would take an act of Congress to disestablish the refuge, and funding cannot be completely cut off as long as the FWS and its refuge system remain in place. Unlike the earlier release experiments, the refuge does not require the actual survival and reproduction of masked bobwhites for its continued existence; bureaucratic capital now flows through the institution of the refuge as a whole rather than through the bodies of individual birds.

The aim of this chapter is to explain these shifts in the form and content of value production on the Buenos Aires. Three dimensions of refuge production are addressed: the management planning and land acquisition efforts launched after creation of the refuge; the activities of masked bobwhite work and prescribed burning; and the role of signs and paperwork in the day-to-day operation of the refuge.

ESTABLISHING THE REFUGE
MANAGEMENT PLANNING AND LAND ACQUISITION

At its most basic level, the creation of the refuge was achieved by a change in ownership: The FWS bought the ranch, removing it from the private economy (map 4). This action followed from the Masked Bobwhite Recovery Plan (Brown and Ellis 1977, revised 1984), but it did not dictate how the refuge would be managed. Immediately after purchasing the ranch, the FWS set out to write a management plan. One was completed in 1987 and another is in production at this writing; both were undertaken after extensive meetings to gather "public input" in keeping with federal laws. No plan for the refuge, however, has ever been approved and implemented. Officially, the 1987 plan was shelved pending resolution of refuge tenure on roughly ninety thousand acres of State Trust lands leased to the former ranch. The refuge held the leases from 1985 to 1991, when a land swap gave the FWS outright ownership.

Examination of the draft management plan and associated docu-

Map 4. Altar Valley circa 1995, showing the boundaries of the Buenos Aires National Wildlife Refuge. The main part of the refuge is nearly identical to the ranch purchased in 1985; the portions in Brown Canyon and along Arivaca Creek were acquired later. (Map by Don Larson)

ments[1] suggests that public use was the primary reason why the FWS wanted a refuge so much larger than the masked bobwhite program required; it also casts doubt on the official justification for shelving the 1987 plan. Researchers had determined that the masked bobwhite program required approximately five thousand acres, less than 5 percent of the overall ranch (Brown and Ellis 1977). A map prepared for the 1987 master plan depicts "suitable/high potential" masked bobwhite habitat as the loamy bottomlands running north-south through the refuge: perhaps a quarter of the whole. Most of this land was deeded, whereas the uplands were predominantly State Trust lands. Bowing to Brenner's demands, the FWS bought the whole ranch, obtaining ownership of the deeded acres and leasehold on the State Trust lands. This posed a difficulty, however. The deeded parcels—more than a dozen of them ranging in size from eighty to more than five thousand acres—were scattered among the leased acres, and the boundaries rarely corresponded with fences, roads, or any conspicuous feature on the land. It would be almost impossible to manage them separately from the State Trust lands and very expensive even to try. The FWS anticipated this problem but failed to recognize the full extent of it.

The State Land Department was willing to transfer the leases to the refuge, but it could not continue to treat them as grazing leases if the new lessee did not graze livestock. By its policies, any nonagricultural lease had to be treated as commercial, and commercial rates were determined by the market value of the land rather than its livestock carrying capacity. Leases that had cost the Victorio Company around twenty thousand dollars per year would have cost the FWS well over half a million dollars annually. When the Arizona State Land Department chose to enforce this policy beginning in early 1987, the FWS's regional director, Michael Spear, had to redirect regional funds while petitioning the Washington office for additional money. The rate was negotiated downward and phased in gradually; all told, the FWS paid the Arizona State Land Department more than three hundred thousand dollars to hold the leases between 1985 and 1991.

In the long term, paying the commercial lease rates was out of the question. Giving up the leases entirely, however, would have left the refuge with a fragmented patchwork of unfenced parcels surrounded by pastures that someone else would likely lease and graze, raising again the management problems that had plagued the reintroduction experiments of the 1970s. Some way had to be found to consolidate the refuge's lands. As it happened, a major land swap between the State Land Department, the FWS, and the U.S. Bureau of Land Management (BLM) was under negotiation, and arrangements were made to include the Buenos Aires in the package: 1,490 acres of high-value, surplus FWS lands at Lake Havasu National Wildlife Refuge would be exchanged for 11,810 acres of lower-value, State Trust lands on the Buenos Aires. These would be selected in such a way as to connect up the parcels already in refuge ownership, concentrating on the area south of Arivaca Road, which was considered the best masked bobwhite habitat. By December 1987, the deal was apparently settled.[2] The planned land swap would have resulted in a refuge of thirty thousand acres.

The FWS had grander ambitions for the Buenos Aires, however. The first draft of the management plan called for extensive habitat manipulation to engineer a bounty of wildlife for public viewing. "Rather than try to reconstruct the original southeastern Arizona ecosystem, this plan provides for altering the existing habitat to encourage numbers of bobwhites and a variety of species which would not have been supported by the original ecosystem" (FWS 1986a: 1). The other species in question were primarily migratory waterfowl, which had been sighted in large numbers on Aguirre Lake in the years leading up to acquisition (see chapter 8). Although in no way in danger of extinction, these birds were designated "species of special emphasis" because of their relative rarity in the state of Arizona. The draft plan called for developing "new or more extensive aquatic and riparian habitats" in order to attract bird species "never before found in the area" and thereby to make the Buenos Aires "more diverse" than ever before (ibid.).

This broader management goal of maximizing "wildlife diversity" served several ends at once. First, it filled a gap created by the uncertainty surrounding masked bobwhite habitat needs. Restoring the Buenos Aires to its "original" Sonoran savanna grassland conditions was the centerpiece of management for the masked bobwhite, but no one really knew what the original conditions were, let alone how to re-create them. Aiming to produce a diversity of habitats sidestepped this problem.

Second, the FWS hoped that a diversity of wildlife, especially rare wildlife, would attract public visitors. Third, the prospect of diverse wildlife *combined* with extensive public use served to justify refuge control of the entire former ranch (FWS 1986b: 4). With enough space, there could be masked bobwhites in the bottomlands, waterfowl at Aguirre Lake, pronghorns in the uplands, endangered fish in the stock tanks, and visitors along old ranch roads admiring it all.

Without modifying the goals of the first draft, subsequent versions of the plan renounced artificial habitat manipulation in favor of "habitats that are natural or native to this area."[3] Habitat manipulation was expensive, and some members of the Master Plan Team voiced concerns that refuge budgets would be inadequate to support so many concurrent programs. These objections were mollified by the notion that "native" habitats would produce themselves: Aguirre Lake would fill on its own, for example, at no cost to management. Subsequent events have disappointed these hopes, but at the time, FWS planners saw no conflict between the masked bobwhite and other wildlife programs. Planners openly expressed ambitions to obtain "management control" not only of the "suitable/high potential" masked bobwhite habitat (most of which the refuge already owned), but also of the "medium potential" habitat, which included the alluvial mesas that composed the majority of the rest of the ranch (FWS 1986b).[4] Outright purchase was a possibility but a very expensive one; easements or some sort of cooperative agreement were considered more viable options.

The land swap plan of December 1987 was inadequate to these loftier goals. The only obstacle to a much larger land swap was the fact that the FWS did not own sufficiently valuable surplus lands in Arizona to trade for the ninety thousand acres of State Trust lands on the Buenos Aires Refuge. The BLM did, however, and a deal was hastily arranged. Regional Director Spear and Arizona BLM Director Dean Bibles conceived a "grandiose plan" to swap the Santa Rita Experimental Range for almost all of the State Trust lands leased to the Buenos Aires Refuge (FWS 1988: 8). Because the experimental range is close to Tucson, its per-acre value was appraised according to its development potential, whereas the more remote refuge land was appraised according to its agricultural value for livestock production. In this way the two were deemed of equal value, despite their significant difference in size (about fifty-six thousand acres and ninety thousand acres, respectively). In the closing weeks of the 100th Congress, the plan was written into the 1988 Idaho-Arizona Conservation Act, a huge public lands bill. The change attracted little notice at the time. The refuge took possession of 89,883 acres of State Trust lands on the Buenos Aires Refuge in April 1991.

The Master Plan Team, composed of representatives from the FWS, the State Land Department, and the AGFD, worked under a cloud of uncertainty generated by the various land-swap plans and negotiations. The FWS's ambitions for public use and land acquisition were not directly challenged in the planning process, but grazing provoked ardent debate. Prior to acquisition, FWS officials had publicly indicated that a study would be conducted to determine if grazing should occur on the refuge (FWS 1984b: 50); a 1986 draft called for a five-year "experimental grazing plan."[5] According to the refuge manager, Regional Director Spear opposed grazing, whereas his assistant, Ellis Klett, favored it. Opinions were strongly divided among planning team members as well. As one member stated in commenting on a draft, "The sentence 'many people feel it would be a "crime" to withdraw this land from its historic use' should be deleted. . . . Many other people feel that it would be a crime to continue with grazing." The final plan was equivocal: "Grazing and woodcutting may be permitted . . . if proven to be beneficial to habitat management" (FWS 1987: 3); the research component was omitted.

The most important outcome of the planning process was not the plan itself but the discovery that managing the Buenos Aires Refuge collaboratively would be difficult. The AGFD had a stake in hunting; until the land swap, the State Land Department retained rights on firewood cutting, beekeeping, and other potential uses of its lands; everyone had an opinion on grazing. Should hunters be allowed to drive off road to retrieve their kills? When a draft plan said "yes," many people responded with a vehement "no." Some praised the drafts as concise and well written; others assailed them for vagueness, misspellings, outdated scientific names, or ecological naïveté. A few team members pointed out that public use plans would be expensive and had nothing to do with the masked bobwhite. The disputes appear to have raised fears within the FWS about their autonomy to make management decisions. The plan was used for the purposes of an environmental assessment required by federal law (FWS 1985), but it was never formally approved as a management plan. Instead, the uncertain status of the State Trust lands was invoked as grounds to shelve the document indefinitely, much to the chagrin of some Master Plan Team members. The grazing experiments have never been conducted.[6]

Several people involved in the 1985–87 planning process believe that the plan was shelved simply because the FWS did not want its "hands tied" with respect to key management issues. According to planning documents, the process assumed "FWS control of all refuge lands" (FWS 1986b: 4). Yet when that control was finally obtained, the plan was not revived, let alone implemented. It is clear, moreover, that the absence of an approved management plan has allowed refuge officials to make decisions more flexibly and opportunistically than would otherwise have been the case. In particular, the importance of the masked bobwhite relative to other management objectives has ebbed and flowed in response to continued frustrations in the reintroduction program, outside criticisms from scientists and cattlemen's associations, budgetary shortages, and media coverage. The land-acquisition program—one of the most aggressive in the FWS—is a case in point. Because funds for day-to-day operations are already inadequate to meet the needs of refuge programs, the relative ease with which new parcels have been acquired has forced the issue of management trade-offs. What is the priority: the masked bobwhite, wildlife in general, the Sonoran savanna grassland, or public use? The official answer is invariably all of them, in that order; in practice the answer has been any of them, according to circumstances. This confused mandate, issuing from multiple laws and conflicting directives from the regional office, is one reason why refuge management has propagated the idea that Nature is in charge on the Buenos Aires Refuge. The other part of the explanation, public use, is the subject of the next chapter.

BIOLOGY AND THE MASKED BOBWHITE
Refuge biologists inherited all of the challenges faced by earlier masked bobwhite researchers except that of cattle grazing. Notwithstanding their predecessors' achievements, they still knew very little about the

bird. An early draft of the first refuge management plan put it bluntly (subsequent drafts were more muted):

> This management plan is by necessity experimental. We are ignorant of the needs of the masked bobwhite. We are ignorant of the response of this type of landscape when grazing ceases to be a dominant impact on the habitat. We are unsure what the original complement of animal species was in this ecosystem. We do not know the carrying capacity of the land for those species which currently exist on the refuge, let alone those which were extirpated but might be reintroduced. (FWS 1986a: 1)

There was a new problem to face, moreover, in the form of raised expectations and scrutiny. Creating the refuge made it necessary to keep up an image for public consumption. In 1985, for example, FWS officials publicly praised the exemplary condition of the ranch, even while they complained in internal memos that Brenner had dramatically damaged masked bobwhite habitat during his two-year tenure.[7]

Until 1996 masked bobwhite propagation was handled at the Patuxent Wildlife Research Center in Maryland, as was described in chapter 1. With the creation of the National Biological Survey, the quail were kicked out of Patuxent and moved to a new facility built on a piece of refuge property southeast of Arivaca. Propagation techniques have been modified slightly (FWS 1995: 59), but they remain basically the same as those perfected in the 1970s.

Mitochondrial DNA analysis has indicated that the breeding population is "genetically healthy" despite generations in captivity (meaning it is not inbred; FWS 1995: 53–54), but the question remains whether the birds retain the necessary instinctual endowment to survive when released to the wild. Cannibalism, disease, and trauma remain the greatest threats to captive birds.[8] Release protocol has been modified more significantly. The surrogate parent methods were changed in the mid-1990s when it was determined that the Texas bobwhite males were abandoning their charges soon after release.[9] At present, grown masked bobwhites are used instead; larger groups of chicks are released with multiple adults rather than only one. Refuge biologists have also manipulated the timing and sequencing of acclimation to the wild: providing more time in flight pens to learn foraging, escape, and covey behavior; holding coveys in temporary pens at release sites for a week before opening the door; and releasing birds at different times of year and at different ages (FWS 1995: 59). According to refuge biologists, abandonment is down and survival rates are higher under this protocol. The twenty-four-day training regimen developed by Ellis has not been continued.

After releasing the masked bobwhites in late summer and early fall, there is virtually nothing that refuge biologists can do except wait and watch, attempting to glean as much information as possible from the birds' actions. Monitoring is done by several means. The most direct method is radio telemetry using very small transmitters affixed around birds' necks (fig. 15). Because of the high cost of the transmitters and the possibility that collared birds are at greater risk of predation, only a small number are monitored in this way. Telemetry is most useful for indicating covey movements in the first few weeks after release; mortality of birds and of batteries makes it rare for a signal to be detectable for more than three months. Live traps are set during the winter and spring months, though much larger numbers of Gambel's and scaled quail are typically caught than masked bobwhites; three hundred Gambel's quail and only four masked bobwhites were caught in 1995 (FWS 1995: 47). Transects are walked, sometimes with dogs, and in the breeding season call counts are performed, in which staff members or volunteers listen for the distinctive "bob-white" call of breeding males. Because all released birds are banded, any found dead or caught in a trap can be individually identified. All random sightings are reported and documented.

In spite of these efforts, the number of masked bobwhites surviving on the Buenos Aires Refuge remains subject to significant uncertainty. Sighting them is difficult: They are extremely reclusive, hiding under thick grasses and shrubs and often not flushing unless nearly stepped on. They tend to avoid traps and open areas such as roads. Females are difficult to distinguish from other quail species, so public sightings are not very reliable. For these reasons, statistical extrapolations of transect and call-count data suffer from small sample sizes and high margins of error.

Figure 15. Refuge biologist using radio telemetry to locate masked bobwhites a few days after they were released from pens, 1997. The embankment on the right is an old stock tank. (Photograph by the author)

The census estimates included in the 1995 Refuge Annual Narrative Report are illustrative. Sixty-three miles of line transects yielded only two coveys of masked bobwhites. The coefficient of variation was estimated at 70 percent, but the results were nevertheless extrapolated to the entire area of "suitable/high habitat" on the refuge (approximately twenty thousand acres), yielding an estimate of 364 individuals. The estimate thus assumes one of the very things that censuses have yet to establish: that masked bobwhites are in fact utilizing all of the habitat designated as highly suitable for them. Bird dog surveys turned up five coveys but yielded an even lower estimate (two hundred) after statistical manipulation. Then, curiously, both estimates were adjusted upward to "a more accurate estimate" of "about 500 birds" (FWS 1995: 46–47). The reasons given for the adjustment were the high coefficient of variation for the methods used (a tautology), poor scenting conditions for dogs, inexperienced dogs, and the fact that biologists "were aware of a number of coveys the dogs never found" (an implicit rejection of the censusing method as a whole).

Because of poor summer rains, 1995 was probably a bad year for the quail; estimates have varied from three hundred to one thousand birds through most of the life of the refuge, during which time about two thousand birds have been released annually. The biologists cannot be faulted for their efforts—there is no better way to count the birds short of stationing an army of observers throughout the refuge—but statistically the estimates are virtually meaningless. This conclusion, however, is disallowed by the bureaucratically and politically contentious circumstances surrounding the program as a whole: The refuge simply must assert knowledge on this point.

From a scientific perspective, the gross number of birds is of less importance than where they are found and why they are there. Population dynamics among masked bobwhites are highly variable. Like other gallinaceous birds, they survive by producing much larger numbers of offspring than eventually breed (meaning that mortality is high even in good circumstances), and total numbers may be expected to vary dramatically from year to year. If one assumes that the birds selectively utilize areas favorable to their reproduction, the challenge becomes how to evaluate such sites: for what are the birds selecting? From the observed behavior of masked bobwhites released on the Buenos Aires Ranch in the late 1970s, Goodwin (1982) developed quantified parameters for suitable habitat measured in terms derived from Stoddard (1931): food and cover. Refuge researchers have extended this analysis to include vegetation structure and microclimatic data (King 1998). The variables are different, but the assumptions of this research are the same as for early range-cattle studies: each observed animal stands for all others; their preferences are seasonal but otherwise static; and habitat manipulations based on the results will improve the animals' survival and growth. In the case of the Buenos Aires Refuge, habitat manipulations are ongoing alongside efforts to understand the masked bobwhite's "preferences." Several tools have been utilized, but the only one applied on a large scale has been prescribed burning of vegetation.

The refuge was created at a time when desert grassland ecologists were increasingly convinced of the historical importance of wildfires. That fire suppression favored woody species over grasses had been noted by David Griffiths (1910) and forcefully argued by Robert Humphrey (1958, 1962), among others. Hastings and Turner (1965), however, had downplayed fire relative to climate as a factor in regional vegetation change. It was only with the work of Henry Dobyns (1981), Stephen Pyne (1982), and Conrad Bahre (1985) that desert grassland fires returned to the scientific spotlight. Dobyns drew on ethnographic evidence to argue that fires had been a regular hunting tool of Apaches, Pimas, and other Native Americans. Pyne echoed this view and extended it to human-fire relations generally. Bahre examined historical documents and concluded that fires had been "fairly frequent" in southern Arizona prior to 1882, when they diminished as a direct result of cattle grazing.

These provocative findings were of little practical use in most of southern Arizona, however, because grass cover was inadequate to carry a fire. Thanks to the Victorio Company's revegetation efforts, the Buenos Aires was an exception. If the masked bobwhite required grasslands, and the grasslands required fire, then fire and the quail should be brought back together. Speaking to a symposium convened to discuss the proposed refuge in December 1984, David Brown stressed, "These grasslands are naturally maintained by fire. In fact, it is particularly essential that grassland bobwhite habitat be maintained by fire. They will not maintain themselves if they are kept free from fire" (Stromberg et al. 1986: 3).

In theory, the fire program is justified by ecological goals intimately tied to the masked bobwhite program. In practice, however, the bureaucratic separation of the two divisions undermines coordination. The refuge bureaucracy is divided into Fire, Biology, Public Use, and Operations Divisions. The Fire Division is the largest in terms of employees, and it enjoys relative autonomy from other divisions for purposes of budgeting. The importance of this in bureaucratic terms cannot be overstated because it alters the relation of the fire management officer (FMO) to the refuge manager and of each to the regional office. In fiscal years 1993 and 1994, when the rest of the refuge underwent budget cuts mandated by

Congress, the Fire Division budget increased slightly over previous years. The Fire Division's supply of labor and equipment is viewed with envy by other refuge staff members, and its employees are routinely called upon to assist in maintenance and other tasks.

The Fire Management Plan divides the refuge into fifty to fifty-five "burn units," the largest of which are three thousand to five thousand acres in size. Burn units may be likened to the pastures of the ranch, except that they are bordered by fire breaks (mostly roads and washes) rather than fences. The plan dictates a seven-year rotation for most units, and it specifies detailed parameters of temperature, humidity, and wind (both direction and velocity) within which each unit may be burned. All other fires must be suppressed unless the same prescriptive conditions are met. According to the FMO, he is *personally* liable if he authorizes a burn outside of these parameters. Ignition can be from hand torches or from airplanes using a device that drops incendiary chemical mixtures in ping-pong-ball–sized projectiles.

The enemies in this war are the "invasive" species: mesquite trees, Lehmann lovegrass, snakeweed, and burroweed. The last two die readily, but the other two are recalcitrant. Seedling mesquites (less than one centimeter in diameter) usually die, as do a small number of larger trees, but most are at best "top killed," meaning that the aboveground portion of the plant dies but not the root system. This damages their ability to set seed for a number of years while the plant recovers. The FMO estimates that in "about three [human] generations" the mesquite population on the Buenos Aires Refuge may be significantly reduced by his division's efforts. The refuge manager asserts that fires are causing Lehmann love-grass to diminish relative to native species, but he has produced no scientific data to support the claim, of which outside ecologists are skeptical (Cable 1965, 1971), and Lehmann lovegrass still dominates large areas of the refuge.

Whether and how the fires affect habitat for masked bobwhites are perhaps the central questions of refuge management at this time, questions that are by no means resolved. Based on casual observations, refuge biologists believe that conditions are best for the birds three to five years after burning, when grasses have had a chance to grow high enough to provide cover (FWS 1995: 42). It is possible, however, that no straightforwardly beneficial relationship exists between fire and the quail. Rigorous, statistical analysis of habitat selection by released masked bobwhites indicates that vegetation structure, especially cover of one to two meters in height, is more significant than composition (King 1998). Low, shrubby mesquites provide such cover (Guthery et al. 2000). The data also suggest that masked bobwhites use areas of exotic grasses disproportionately *more* than native grass stands, perhaps for reasons relating to cover (King 1998). Biologists cannot know, however, if the birds "choose" the non-native grasses or use them for lack of adequate stands of native grass. In short, the ecological relationships between the masked bobwhite and the habitat produced on the Buenos Aires Refuge are still not well understood. The hypothesis—and it always was just a hypothesis—of a "natural" affinity among the bird, fire, and native grasses has yet to be proved or disproved.

THE REFUGE'S RELATION TO NATURE

The activities of the Biology and Fire Divisions reveal in detail the FWS's relations to nature on the Buenos Aires Refuge. The masked bobwhite and its habitat are manipulated and controlled as much as cattle and range were before, arguably more so, with technical capacities weighed against monetary resources to determine what actions to take. But the boundaries between what is controlled and what is not are marked differently than on the ranch. To a rancher, uncertainty amounts to a risk that may threaten continued operation of the ranch. Nature is controlled if possible, submitted to if not; like fortune, it is simultaneously ally and adversary, succor and threat (Pitkin 1984). In contrast, the refuge does not directly depend on the survival and reproduction of masked bobwhites for its institutional continuation. Within budgetary constraints, whatever can be controlled by the application of labor and technology is controlled. Those things that cannot be controlled, however, do not represent a risk; to the contrary, they are valorized as "natural." Thus, captive birds are bred in accordance with a strategy of optimization using commercially derived techniques of total environmental control, resulting in unnaturally large numbers of offspring. Once released to the wild, however, the birds are out of human control and begin the journey back to the status of "wild," or so it is hoped. One refuge biologist justifies the abandonment of Ellis's training regimen on the grounds that only "natural" selection should operate in determining which birds survive postrelease. The goal of the program is a "self-sustaining population," meaning that "nature" would take over and releases would no longer be required.[10]

The fire program demonstrates that the fundamental criterion is instrumental control rather than "naturalness." Wildfires must be suppressed, even as substantial resources are devoted to burning the area on purpose, because they defy the parameters established by human calculations of what is desirable and because they can be controlled. That fires are "natural" and grubbing and chaining are not is a secondary rationalization: Management openly admits that the more intensive methods are simply too expensive given their budget.

Like the ranch, the refuge must ultimately rely on natural processes for its production (although not for its reproduction); in both cases "nature" refers to this dependence, that is, to those factors that are beyond instrumental control. If, as Lawrence (1982) argues, ranching strives to dominate nature, the same must be said for the refuge. Lawrence's formulation, however, reifies nature, obscuring the ways in which its boundaries shift in response to economic, technical, and social conditions. The refuge approaches the natural world through the tools and concepts of science, but insofar as science enables instrumental control, nature becomes a category of what we do *not* know. Is it sound to take released masked bobwhites' "preferences" as "natural" when there is doubt as to whether they still "know" how to survive in "the wild"? Existing endangered species only invert the problem. In 1989, for example, the Pima pineapple cactus *(Coryphantha sheeri robustispina),* a listed endangered subspecies, was found on the refuge, and the potential threat posed to it by the fire program caused several years of disruption. The Ecological Services Division of the FWS ordered that the cacti be protected by careful advance burning of areas around them, and hundreds of hours of labor were devoted to inspecting burn units on foot: "61 staff hours for each cactus located" in 1995 (FWS 1995: 30). Presumably, either the cactus is adapted to withstand fire or it did not "originally" occupy the area, but a boundary was drawn nonetheless, bureaucratically and on the ground: The cactus is part of nature and must not be harmed by official refuge actions. Meanwhile, studies were launched to determine if the cactus could be grown under controlled conditions; if so, restrictions "in the wild" may be relaxed.

The ambition to restore the Buenos Aires Refuge to its "original" condition of Sonoran savanna grassland was a logical extension of masked bobwhite reintroduction. The need to erase human agency expanded from the bird to encompass the entire refuge, envisioned as a state of nature: a self-sustaining system of vegetation, soils, water, and wildlife. From the perspective of scientific ecology, restoration was an impossible (if not meaningless) objective: No one knows what the Buenos Aires Refuge "originally" looked like, and the clock cannot be turned back in any event. The bureaucratic mode of reproduction favored symbolism over science, however, opening a gap between events on the ground and their official representations. The continued frustrations of the masked bobwhite program have only widened this gap. But production and reproduction cannot be entirely disjoined. How are they related? Here we must turn away from ecological questions and face the social ones directly.

SIGNS OF THE STATE

Upon acquiring the Buenos Aires, the FWS promptly set about installing signs. Along the Sasabe highway at the north and south boundaries and at the mouth of the main driveway, large signs announced "BUENOS AIRES NATIONAL WILDLIFE REFUGE." Smaller signs were placed every quarter mile around the perimeter of the refuge and on both sides of the two paved highways cutting through it. Speed limit signs appeared along the main driveway. Other signs followed over time, designating areas for public travel, hunting, camping, wildlife viewing, and so forth. This abundance of signage was the most obvious phenomenological marker of the land's new status. It is safe to say that no previous owner had ever seen the need for so many signs on and around the Buenos Aires. Yet the signs of the refuge are taken for granted by almost everyone who works or visits there.

Only local residents have taken issue with refuge signs, and only with some of them. The perimeter signs, which say "NATIONAL WILDLIFE REFUGE—UNAUTHORIZED ENTRY PROHIBITED," have raised objections because the refuge is in fact open to public use, and the signs therefore appear both misleading and intimidating. The large entry sign erected near the Arivaca Ciénega in the early 1990s has been vandalized twice, once by arson and once by theft. No one was prosecuted, but refuge officials and local residents viewed the incidents as acts of opposition perpetrated by local miscreants. Locals would agree, I think, that the signs merit attention, not only for what they represent but also for what they *do*. I do not claim to articulate local views in this section, however. I wish to pursue a more theoretical train of thought, using the signs themselves as ethnographic data.

Next to major highways in Mexico, usually as one leaves an urban area, are found red and white road signs that carry a rather peculiar message. "OBEDEZCA LOS SEÑALES," they say: "Obey road signs." In the United States, apparently, there is no need for such reminders. As the U.S. Department of Transportation notes confidently, "Signs ordinarily are not needed to confirm rules of the road" (1978: 2A-1). After all, if the other road signs are not already obeyed, then why would people obey this one? Regress ensues: Each sign requires another to affirm its authority, and so on. The Mexican signs disrupt and thereby expose the assumptions made by road signs everywhere: that observers will read and understand the messages and that the messages so conveyed will affect their behavior.

It is the taken-for-granted, self-positing authority of road signs that makes them particularly representative of the workings of the state (Bourdieu 1998: 51). Like myth for Barthes (1957), road signs are second-order semiological systems: The signifiers are the signs themselves, and the signifieds are people's actions, which the signs are intended to affect. But what is the signs' relation to social actions? How do signs work, or more precisely, what work do they do? A few examples from the Buenos Aires Refuge may elucidate some answers.

The sign pictured in figure 16 says that the speed limit on the refuge driveway is fifteen miles per hour. This is an utterance of the state, and what it utters is the law. The sign creates a space within which the law *is* what the sign *says*. Of course there are places where the speed limit is not entirely clear, but such signs largely succeed in establishing a specific relationship between people, space, and the state, with the appearance of universalist objectivity: fixed, visible, and applicable to all. Indeed, uni-

Figure 16. Signs installed by refuge workers along the main road into the head-quarters, 1997. (Photograph by the author)

formity is the central organizing principle of traffic control signage (U.S. Department of Transportation 1978: 1A-3). It is this relationship that was absent on this road before the refuge was created.

If people see this sign and drive no more than fifteen miles per hour, we might say that they obey the sign. Two reasons could be given in explanation. First, they wish to avoid having an accident, and they see the sign as a guide to their own safety. As the people of Montana would no doubt point out, this fails to explain why the sign is necessary in the first place. The need for the sign must reside elsewhere; or rather, the reader of the sign must reside elsewhere, unfamiliarity with local conditions becoming the foremost justification for the sign. This assumes, however, that the sign is correct—that driving more than fifteen miles per hour is unsafe on this road—which I can report from experience is not the case. Second, we might invoke a more coercive explanation: Drivers obey because they fear the repercussions of breaking the law. Driving twenty-five miles per hour may be safe, but it is nevertheless illegal. The problem with these rather functionalist arguments is that they are both misleading and beside the point. They are misleading because they ascribe conscious rationality to a context deeply determined by prereflexive, embodied habits. To explain driving in terms of road signs is like explaining speaking in terms of rules (Bourdieu 1977: 30). Even the Federal Highway Administration recognizes this; it is a principal reason for uniformity in traffic control devices (U.S. Department of Transportation 1978: 1A-3). The arguments are beside the point because no one obeys this speed limit, not even the refuge officials empowered to enforce it.

If people exceed the speed limit but do not get caught, what happens? Where are all those unrecorded incidents of speeding? Officially they do not exist. As the medium of rule of the state, documentation is more than an epistemology. In practice, it becomes an ontology: That which is documented, is; that which is not documented, is not. When a refuge employee intentionally ran over a coyote pup on the way to work, for example, the refuge manager declined at first to write him up, exercising his power to decide if the event would exist in the bureaucratic sense. Only after animal rights' activists complained did it result in an official representation, a permanent mark on the employee's record. Any attempt to study the state scientifically must grapple with the distorted duality of the unrecorded and recorded worlds.

The signs in figure 17 are located at the entrance to Brown Canyon, a site one might describe as a situation of stressed signification. The signs concern hunting; they are on the refuge but were erected by the AGFD. As in the case of the speed limit, the signs here attempt to define what people may and may not do, but in this case the spatial coordinates are much more complex. Despite a map, taped to the inside of the lid of the white box in the picture, many visitors remain confused about the exact boundaries between State Trust land, where hunting is allowed, and private or refuge land in Brown Canyon, where it is not. This greater complexity and confusion forces these signs to be more verbose than in our previous example.

Imagine that you are a hunter arriving at this site. The closed gate across the road compels you to stop and get out of your vehicle, making it hard to avoid the signs. Most hunters, I expect, scarcely read the signs at

Figure 17. Signs at the gate into Brown Canyon, 1997. (Photograph by the author)

all, being already familiar with the protocol from prior experience. In the lower left, the smallest sign carries an explicitly legal message: "PLEASE SIGN IN/OUT HERE. Failure to sign in prior to entering onto private property is considered trespassing and may be subject to trespass laws." The relevant sections of the Arizona Revised Statutes are then referenced, as if to say "You can look it up." Presumption is here reversed: You must document your actions, or they will be automatically in violation of the rules. The sign is printed in red on white, the color scheme of prohibition in standardized public signage.

The sign in the upper left, printed in black, contains a list of actions that are recommended but not backed by the same threat of law. It speaks to your relationship to the private property owner across whose land you are about to pass. The owner permits your passage but may revoke that permission "at any time"; failure to abide by the suggested guidelines, you infer, may lead to revocation. Phone numbers are provided to report vandalism or violations of game laws. The sign to the right informs you that the system of regulations is not only spatial but temporal as well. Under the aegis of the "Landowner-Lessee-Sportsman Relations Committee," the area is closed to vehicular access when big game are not in season.

Together these signs address the contentious and complex relations of public and private property among hunters, ranchers, and the state. Land ownership is clear on paper but often indiscernible on the ground because many boundaries are neither fenced nor marked. Roads pass across different ownership types, meaning that what you may and may not do changes as you move across the landscape. Wildlife likewise moves indiscriminately across ownership boundaries. Because wildlife is public property, citizens have a right to access it for hunting; the state manages the hunt to ensure that this public good is not depleted. To this end, surveys of deer and javelina populations are made annually from helicopters. The state is parceled into game management units, which overlay both private and public lands; each unit is assigned a certain number of permits according to game populations.

A functionalist interpretation of these signs is more helpful than in the previous case, but it is prone to diminishing returns. Regulating the relationship of hunters, ranchers, three agencies of government, and several varieties of wildlife requires careful coordination, which in turn necessitates signs to define the spatiotemporal limits of permitted activities and to inform people of what they may and may not do. The possibility that people may unintentionally go astray from these guidelines motivates the placement and complexity of the signs. The hunter must be made the agent of his or her acts in order to be held responsible for them; the signs produce the individual visitor as a rational subject in relation to a system of spatiotemporally defined laws. In theory, several fields of activity coexist, each with precisely defined spatial and temporal coordinates. In practice, however, conflicts and misunderstandings occur. A vicious cycle may result: Increasing complexity provokes ever more signs, which, in their attempt to explain the system, may obscure it further. The subject aborts, pleading incomprehension. "Care should be taken not to install too many signs. A conservative use of regulatory and warning signs is recommended as these signs, if used to excess, tend to lose their effectiveness" (U.S. Department of Transportation 1978: 2A-3). Taken to its logical conclusion, the functionalist interpretation must concede, paradoxically, that signs may be both symptoms and causes of *dys*function.

So why did the FWS erect so many signs on the Buenos Aires Refuge? Taken individually, each sign appears to serve some practical function, establishing boundaries between prescribed and proscribed activities, state and subject, public and private. Functionalist explanations break down, however, when they are applied to the signs in aggregate or extended to their logical limits. For the employees who put them up, the signs reflect "standard operating procedure," their size, design, and location regulated by policies that apply throughout the FWS. To understand the signs, we must look inside the bureaucratic processes that produced them.

PAPERWORK AND BUREAUCRATIC COMPETITION

Reproduction of the Buenos Aires Ranch occurred by cycling profits (from cattle or some other commodity production) back into the ranch. With the advent of the refuge, the circuit of reproduction no longer flows through commodity exchange. Instead the refuge secures funds through the FWS bureaucracy, a system mediated by money (flowing down from Congress) and paperwork (flowing both directions but especially up).[11] All of the components of refuge production must find paper representation (in plans, maps, reports, surveys, etc.) because refuge reproduction depends on bureaucratic administration. Paperwork must be produced and managed in such a way as to fulfill the norms of the FWS and thereby ensure the flow of funds necessary for continued refuge operation.[12]

Just as the Endangered Species Act made a scientific epistemology into an ontology of government conservation, the creation of the Buenos Aires Refuge imposed state forms of classification on its land, vegetation, wildlife, water, infrastructure, visitors, and employees. There are official policies and procedures to cover virtually every aspect of refuge work: the Masked Bobwhite Recovery Plans (periodically revised), the Fire Management Plan, the Sign Manual, the Personnel Manual, and specific plans for hunting, predator control, trapping, and the herd of reintroduced pronghorns. Every supervisor on the refuge has shelves of documents like these, which they may have read once but infrequently consult. Likewise, every employee has an official job description that may or may not accurately reflect what s/he does. Those things that must be done "by the book" have usually been learned to the point that "the book" is scarcely necessary; most procedures afford at least some flexibility. Quite apart from guiding employees' actions, paperwork does something simply by existing: It asserts the supremacy of the "official" version of things over the reality encountered every day. One example will serve to illuminate several points.

I first encountered Form DI-134, "Report of Accident/Incident," after a visitor to Brown Canyon tripped and broke her ankle. I completed the much simpler FWS "Incident Report" form, later learning that, because of the gravity of the injury (measured, I believe, by the fact that she required hospital treatment), the longer DI-134 form was necessary. This form assigns two-digit codes to fourteen aspects of the accident being reported, with from five to fifty-seven possible codes for each. Under "Type of Accident/Incident" are found codes for "Struck Against; Struck By; Fall From Different Level; Fall on Same Level; Slip or Twist (Not Fall); Caught In, Under or Between; Rubbed or Abraded; Bodily Reaction; Overexertion; Drowning; Contact With Electric Current; Contact With Temperature Extremes; Contact With Radiations, Caustics, Toxic and Noxious Substances; Noise Exposure; Occupational Disease; Bite (Animal, Insect, Etc.); Explosion; Fire; Immersion." Four categories of "Motor Vehicle Accidents" are also listed, as well as two residual categories: "Accident Type Not Elsewhere Identified" and "Unclassified or Insufficient Data." Among the fifty-seven options under "Source" are "Aircraft; Dusts; Ladders (Fixed); Ladders (Portable); Infectious and Parasitic Agents; Particles; Pumps and Prime Movers; Water; Human Being." It was not an option to describe the visitor's accident in its particularity (tripped over rock). I had to think of it in terms of these state forms of classification.

As with the signs examined above, Form DI-134 seeks to establish a spatiotemporal order characterized by stable representations, accountability, and legibility. Whereas the signs seek to regulate actions before they go astray, the form works to contain an event that has already occurred. Accidents are by definition exceptional, however, and most paperwork on the refuge seeks, as do the signs, to head off mishaps before the fact. The act of recording creates a second order of signification in which the completed form is a sign in itself, with a signifier (the act of completing the form) and a signified (obedience to the established procedures of the system represented by the form); hence, "We were just doing our jobs." Converting the unpredictable place-, time-, and person-specific conjuncture of factors that make up human practice into an orderly, manageable system requires this second order of signification. The paperwork and the signs do the same work.

The correspondence between actuality and record, actions and paperwork, is the axiomatic foundation of bureaucratic organization. Once recorded and filed, a piece of paperwork is presumptively accurate; its veracity can only be challenged by recourse to other documents (or documented testimony) that contradict it. A verbal statement, if not congealed in written form, can be reneged, discredited, or disclaimed with relative ease. (Chronic bad relations between the refuge and area residents can be traced in part to the fact that locals value spoken statements and commitments as highly as written ones, whereas the refuge manager does not.) In practice, however, paperwork is "fudged" all the time. It does not take long to learn that completing one's paperwork is more important than accurately representing reality because it is the paperwork that mediates between levels of the bureaucracy. An incomplete form cannot be processed smoothly and is returned for further work; a complete but somewhat inaccurate one is unlikely to be noticed. This is largely a function of scale: National administration is feasible only if events are reduced to a fixed, standardized medium, with local particularities abstracted away.

All money expenditures are subject to particularly detailed paper work representations. Every employee is presumed to be tempted to turn public resources to private advantage. According to official procedures, no employee may purchase anything without completing a series of requests and approvals moving through the chain of command. In my experience on the refuge, the rules were so cumbersome as to make following them impractical. Expensive items, which were typically anticipated, could be properly documented and authorized in advance. Smaller items, however, for which the need often arose suddenly, would routinely be authorized over the phone and the paperwork completed after the fact.

More complex situations provoked ingenious evasions. The aim was not to defraud the FWS but quite the opposite: to accomplish basic and necessary tasks. In this way the practical necessities of reality were accommodated by permitting a margin of divergence between reality and paperwork. It was the paperwork, after all, that had to be in order.

The structure of bureaucratic production is thus prone to a vicious cycle like the one encountered in the case of signs. Sooner or later a discrepancy is noticed: some form that is inaccurate or a gap in the record where paperwork is missing. This may lead to a search for similar occurrences, and a pattern may be discovered. The supervisory layer of the bureaucracy reacts by mandating a new procedure to follow or form to fill out. In the fall of 1996, for example, a suspicious congressional committee demanded an accounting of every hour of every FWS employee's time in the fiscal year just ending. The paperwork that resulted could only be the wildest approximation because records quite simply had not been kept for that purpose. Nevertheless, the documentation was quickly produced. (Some FWS employees remarked that they were actually lucky: USFS employees had to account for every *quarter* hour.) It is likely that the errors or omissions reflected an excess, rather than a lack, of paperwork and oversight and that the new requirements merely exacerbated the problem. As new policies and paperwork are added atop layers of the old, the only measure of success remains the same: the continued smooth operation of the gears of bureaucratic organization. No matter how precisely the paperwork attempts to regulate people's actions, some room for discretion remains; if it did not, there would scarcely be a need for human employees.

The structure of the bureaucracy simultaneously constrains and empowers those who occupy its offices as they choose what to record and how, when to bend the rules and when not to, alternately invoking and ignoring the official procedures and hierarchies in order to advance institutional and individual agendas (Sewell 1992). The structure does not rigidly determine the outcome or favor any particular strategy, however. Two maxims cited by bureaucrats in the FWS and other government agencies are illuminating in this regard. One is simply "CYA: Cover your ass," which refers to the need to keep careful records of events in order to have documentation, if necessary, for future reference. This maxim applies equally to events working for or against an employee, whether in relation to superiors, subordinates, or representatives of other agencies or entities. Generally, CYA applies most forcefully to subordinates, who are the least likely to be believed in the absence of a paper record. But because most employees are superior to some and subordinate to others, the application of CYA is selective. For example, the Buenos Aires Refuge manager, in his handling of disciplinary issues within the refuge, has shown a tendency *not* to document incidents according to official procedures. By leaving things off the record, he keeps them between himself and the individual involved, perhaps as a favor to be redeemed in the future, as a form of damage control (preventing *his* superiors from knowing of the incident), or simply to avoid the hassle. The second maxim—"Ask forgiveness not permission"—displays a similar logic. It refers to the fact that superiors are unlikely to approve an exception to policy or procedure if asked ahead of time but are likely to let it go if presented with it after the fact. The phrase is invoked when doing "the right thing" involves bending the rules in some way, with the understanding that a recognizably positive outcome will override the procedural misdemeanor in the superior's judgment.

It is in this context that bureaucratic competition takes place. It involves the official hierarchy itself, within which each employee's place is explicitly marked by job title and government service (GS) grade: the refuge manager is a GS-13, the senior biologist is a GS-12, the other division heads are GS-11s, and so on. Supervisory authority follows vertical lines within the organizational pyramid, and no one may be the supervisor of a person of equal or higher grade. (There are further gradations within each GS level, but these are principally steps in each employee's career ladder.) Competition operates along recursive and segmentary principles, the bureaucratic hierarchy resembling a huge, synchronic lineage structure: Each employee competes with those laterally related through a common supervisor. The Biology, Operations, and Public Use Divisions (and to a lesser extent the Fire Division) compete with each other for the refuge manager's approval; these divisions are united, however, along with the refuge manager, in competing with other refuges for the approval of his superiors in the regional office. The regional office, in turn, competes with other regions for the approval of the national office in Washington, D.C.

Although competition is measured in terms of budgets, it is mediated by paperwork and is therefore subject to the dualist ontology described above. When Congress provisionally authorized extra funds for the FWS for fiscal year 1999, for example, the national office instructed the regions to submit lists of projects for the additional funding. Region 2's list turned out to be the smallest in dollar terms. Upon learning this, the deputy regional director instructed all staff to revise their lists upward, adding projects as necessary, in order to ensure a "fair" slice of the pie. This, in turn, prompted a meeting at the refuge, where a wish list of staffing needs, facilities improvements, and other projects was crafted. Actual needs or accomplishments can become subordinate to their paper representations; the farther up the hierarchy one moves—and, ipso facto, the greater the scale of the bureaucracy—the more likely this becomes.

The defining challenge of refuge work is to mediate between paperwork and reality, official and actual, the desk and the field. In theory, these dualities are resolved in and through the work of refuge employees, who produce both the actual outcomes in the field and the paper representations thereof. In practice, however, unequal and competitive bureaucratic relations—both externally and internally and above all with respect to money—tend to pull them apart, at least on the Buenos Aires Refuge. This should not be taken as an anomaly but as a structural feature of the scale of bureaucratic mediation itself. In the absence of an obvious "bottom line," success sometimes means little more than satisfying the potentially capricious and occasionally contradictory expectations of one's superiors. As their representations pass through a national system of offices, actual results may become secondary to diffuse, ungrounded perceptions that are highly susceptible to myth and scandal. If everything were always done strictly by the official procedures, the entire system would probably come to a screeching halt (Scott 1998). The point, therefore, is not that paperwork and reality diverge from each other, although they often do, or that one is essentially more fundamental than the other. Both are equally important to understanding refuge production.

CONCLUSION

Purchasing the Buenos Aires Ranch and transforming it into a national wildlife refuge did not resolve the tensions between science and symbolism that had characterized masked bobwhite reintroduction efforts in the

1970s. On the contrary, acquisition heightened and expanded the contradictions of the earlier period. It rendered the FWS responsible, at least symbolically, not only for the masked bobwhite but also for all other wildlife on the Buenos Aires Refuge. Since 1985 three other threatened or endangered species have been introduced to the refuge (Chihuahuan pronghorns, Chiricahua leopard frogs, and razorback suckers), and dozens of other birds, mammals, reptiles, and plants have been singled out for study and protection. Set into bureaucratic context, reintroductions all pose the challenge examined in the last chapter—how to *produce* something "natural"—whereas the other species present the same problem, only the other way around: How can the FWS claim credit for species whose presence predates the creation of the refuge?

The symbolic politics surrounding the refuge undermine the conditions necessary for effective management: namely, a clear mandate coupled with a certain degree of professional autonomy and integrity. Just as paperwork produces a distorted ontology within the FWS bureaucracy, the need for public support generates a split between the refuge as it actually exists and as it appears in public consciousness. This is characteristic of most government bureaucracies, but it is compounded by a further paradox in the ecological processes "managed" by the refuge. Many of these processes are poorly understood. Even fairly simple questions such as "How many species of birds are found on the refuge?" are subject to a difficulty: how to know for certain. "Nature" becomes something of a black box, with different parties claiming to know what is inside. The various indices (deer, quail, other birds, etc.) used to measure the refuge's performance are supposed to be natural, yet their status depends on human counting and management. Although the masked bobwhite and other valued wildlife are not commodities in the classical sense, they are nonetheless bearers of value that the refuge seeks to maximize; they are the currency of bureaucratic capital accumulation within the FWS. Acts of counting, studying, and representing "official" nature are speculative endeavors in both senses of the term.

We are now in a position to recognize that the definition of "nature," in the context of state conservation agencies, is not fundamentally about ecology. It is about property, specifically public property. The liberal tradition, preoccupied with the individual, ascribes primacy to private property and understands public property as derivative. This is consistent with the genealogy of public lands and wildlife law in the American West: Only after private appropriation had created environmental crises were public land–and wildlife-management agencies created. They were founded on liberal premises: The FWS exists (as do State Game and Fish Departments) because wildlife is legally classified as a form of property. The agency's function is to defend its property interests much as a private property holder would. We will see in chapter 8 that feral animals threaten the very premise on which the FWS rests: namely, that wild is to domestic as public is to private. The possibility that the masked bobwhite is now effectively domesticated—unable to survive in the wild but prolific in captivity—is but the other side of the same coin.

It is in this connection that we may understand the curious fact that the refuge, although public, is itself internally divided into private and public spheres. Most of its decisions and actions are open to public scrutiny, but certain issues, especially those related to personnel matters, are not. The boundary between public and private is strongly marked and enforced in the day-to-day activities of refuge employees. They wear standardized uniforms that signify their status as public employees; certain rules apply only when in uniform or when using a government vehicle. Their days are divided into hours "at work," when they act as public officials, and hours when they are private citizens. Most of them live in government-owned housing on the refuge, but their homes are understood as private spaces. Here again the issue is one of property: namely, the FWS's property right to the labor of its employees. Likewise, elaborate mechanisms exist to prevent employees from taking private advantage of public resources. Refuge property is tagged, numbered, and catalogued, as are the masked bobwhites; refuge money circulates differently than does private money. The opposition between public and private is at once presupposed and reproduced by these social practices. It is this inscribed distinction, more than the actual practices of refuge employees, that differentiates the mode of production of the refuge from that of the ranch. The exception proves the rule: The FMO is held personally liable if he fails to follow official fire parameters to ensure that he conforms to the rules.

In its organization and production, the refuge is entirely determined

by the legal fiction that wildlife is a form of property that can be defended and manipulated on the model of private property. But liberal assumptions founder when applied outside the realm of simple commodity production, as Durkheim (1957) pointed out a century ago. Wildlife is not the product of human labor (at least no one can be identified as its producer); like the grass of the open range period, it is a product of nature. The Buenos Aires and the masked bobwhite have been intensively manipulated and changed from their "original" conditions, but the FWS constructs them as "natural" nonetheless. This idea more than any other has garnered the refuge political support.

To produce wildlife as *public* property is a contradictory endeavor. Without embracing Durkheim's evolutionist framework, we can recognize that public lands and wildlife in the United States have long represented another mode of value from that of private property. Like commodity value, it is at once revealed and obscured by its form of appearance: in this case, as the private property of public agencies. Just as America's national parks have served as pilgrimage sites (Pomeroy 1957) and as vessels of nationalist sentiment (Runte 1979), the Buenos Aires Refuge represents a convergence of nature, the state, and the sacred. Herein lies the importance of public use, the subject of the next chapter.

"where wildlife comes naturally!"

National wildlife refuges have long suffered from a contradiction. Obtaining funds from Congress requires public support, and raising public support usually entails increasing public use. Public use, however, can jeopardize the very wildlife that refuges are supposed to protect. The Buenos Aires Refuge represents the latest incarnation of this well-established pattern.

Refuges came into existence in the early decades of the twentieth century, first by executive order and subsequently by acts of Congress as well. The earliest refuges were small areas where migratory birds bred or fed, and management consisted simply of prohibiting human incursions, especially hunting. As the system expanded to include larger areas set aside for other wildlife, pressure for public access increased, especially for hunting. As early as the 1920s the "sanctuary" philosophy began to erode as compromises were made to secure additional funding and growth (Curtin 1993). The Duck Stamp Act of 1934, for example, created a reliable source of funds for wildlife management, but it also gave hunters greater leverage to demand access to refuges. After World War II, rapidly increasing recreational demand prompted Congress to open refuges to hunting, camping, boating, and other public uses deemed "compatible with the primary purposes for which the area was established" (Refuge Recreation Act of 1962, cited in Curtin 1993: 34). This "doctrine of compatibility" afforded the U.S. Secretary of the Interior and individual refuge managers considerable discretion, which in practice exposed refuges to political pressures of all sorts. Allowing public uses was usually popular, whereas attempting to curtail them generally was not.

By the late 1970s, compatibility had devolved into what one congressional committee described as an "overall policy of 'expanded economic and public use'" on refuges, notwithstanding FWS claims that wildlife remained the highest priority (Curtin 1993: 36). Wildlife advocates decried the meager funding, low morale, and dilapidated facilities at refuges (Defenders of Wildlife 1978; Fischer 1977). Budget cuts in the early 1990s provoked complaints that the refuge system was suffering "an identity crisis" attributable to a lack of public awareness, inadequate budgets, and the absence of a clear legislative mandate. A feature article in *National Geographic* summarized the situation this way: "Over the years Congress has allowed so many different uses of this public land that some critics fear parts of the system have become too public. Grazed, farmed, drilled, logged, contaminated, and open to various kinds of motorized recreation, many refuges now seem in danger of becoming like every other realm where wild creatures give way before the needs of humans" (Chadwick 1996: 11).

The FWS's response to this "crisis" was a publicity campaign built around the slogan "Where wildlife comes naturally!" In 1997 organic legislation was finally passed to guide refuge management nationwide. The National Wildlife Refuge System Improvement Act, however, largely ratified the status quo by recognizing six "priority public uses," again limited only by "compatibility," and it instructed the FWS to cooperate with local citizens and agencies, which could further compromise the ability of managers to put wildlife ahead of human constituencies.

The decision to exclude grazing and promote bird-watching on the Buenos Aires Refuge reflects a shift in the balance of power among competing constituencies. During the 1970s three issues had converged to make grazing on national wildlife refuges a priority target for environmental activists. First, government predator control programs, which had long since eliminated wolves and grizzly bears in the Southwest and continued a genocidal battle on mountain lions and coyotes, drew attention to the complicit relation of government agencies and livestock interests in directly killing wildlife, particularly in Arizona, one of the last states to rescind the bounty on mountain lions (Davis 1970; Johnson 1973a, 1973b, 1974; Murray 1973). Second, the prevalence of hunting on refuges led to increased scrutiny of the FWS as a whole. A survey conducted by Defenders of Wildlife revealed low field morale, top-heavy administration, and a pattern of refuges beholden to sportsmen, ranchers, and recreationalists (Cooper 1978; Defenders of Wildlife 1978; Fischer 1977; Scott 1978; Smith 1982). Third, public lands grazing emerged as an issue in its own right. Ranching was seen to be hurting wildlife not only through predator control but also by degrading habitat on rangelands.

Arguing on behalf of wildlife, activists found that they could press their demands most forcefully on refuges, the only federal lands officially managed for wildlife (Ferguson and Ferguson 1984; Gallizioli 1976; Johnson 1978a, 1978b, 1985). They maintained that refuges should be sanctuaries, free from "consumptive uses" such as hunting, fishing, grazing, farming, mining, and oil drilling. "Nonconsumptive uses"—that is, all types of recreation except hunting—came to appear as the virtuous opposite of consumptive ones. Bird-watching and hiking represented health, nature appreciation, and edifying leisure, whereas hunting was portrayed as killing for pleasure, and grazing, farming, and mining appeared as ways of exploiting nature for profit. The Buenos Aires Refuge owed its creation in significant part to the lobbying of the National Audubon Society, which had prevailed in Congress against opposition from cattle growers' associations.[1] In view of these factors, it is no surprise that the FWS has promoted bird-watching, downplayed hunting, and refused grazing on the Buenos Aires Refuge.

However benign nonconsumptive uses may be compared to consumptive ones, they confront the FWS with a dilemma. "Wildlife values" remain paramount in official statements of management, but it is no secret that increased funding is a high priority for most refuges, including the Buenos Aires. Many FWS employees admit privately that refuges with high rates of public visitation are favored in budgeting decisions. They point to the Bosque del Apache Refuge in New Mexico, where more than one hundred thousand visitors per year have helped to secure the patronage of a powerful senator. Developing and promoting public-use programs and facilities cost money, so management must invest in public use before reaping its rewards. The first Buenos Aires management plan approached this challenge much like a private firm marketing a product:

> Uses encouraged on the refuge will fill a recreational niche left empty by other entities in the area. . . . Birding will probably be the most highly sought activity at Buenos Aires NWR. Prime birding areas will be preserved and open to the public. Normal refuge offerings such as auto tour routes, foot trails, interpretation, photography, etc., will

also support birding. These activities will be concentrated near (but not limited to) the refuge headquarters, where developments will create the greatest diversity and concentration of wildlife. . . . Based on estimates of demand for wildlife and wildlands related activities, this plan aims to accommodate roughly 39,000 visitors annually within the 10 year planning time frame. (FWS 1986b: 3–4)

Just as the captive-bred masked bobwhites had to be thrown into nature to realize their symbolic value, wildlife species on the refuge realize maximum bureaucratic returns only if they are seen by human visitors. Here a further contradiction arises, however: Most wildlife species tend to avoid humans, so efforts must be made to facilitate viewing through signs, brochures, and infrastructure (roads, trails, observation areas, etc.). Yet at the same time, wildlife sightings must appear "natural," spontaneous, and a product of good fortune rather than design because this is what distinguishes a refuge from a zoo. The promotional slogan for the refuge system captures this contradictory imperative perfectly.

In the first part of this chapter, I examine the investments that refuge management has made in promoting public use on the Buenos Aires Refuge. Building on the arguments of chapter 7, I show that public use has been a constant but elusive goal of the refuge from its inception. Promoting the idea of the refuge as pristine or original nature has been the central theme of these efforts, aimed at an audience that, at least in the judgment of the FWS, strongly opposes grazing. By inadvertently drying up Aguirre Lake and other ranch-era water sources, however, refuge management has eliminated the very attractions that early planners placed at the center of public use development. This has led refuge officials to shift their plans to newly acquired lands with more reliable water. Nevertheless, visitation has never met anticipated levels, a failing that refuge management attributes to inadequate funding.

In the second part of this chapter, I turn to public visitors themselves to consider how they construct value through viewing certain animals "in the wild." *Seeing* is a recurrent theme throughout the chapter—wildlife sightings, scenic views, overlooks, etc.—and in this connection the refuge suffers from a curious predicament. The masked bobwhite is extremely difficult to see. Some visitors have complained about this, and the refuge, having long resisted on various grounds, has recently put a few live birds on display. In addition to this physical difficulty there is a socially constructed one: For serious bird-watchers, the masked bob-whites released on the refuge "do not count" as legitimate sightings precisely because they have been reintroduced by human efforts. For a significant portion of the refuge's target audience, its founding attraction is not much of an attraction at all. This combination of factors has forced the refuge to promote public use in terms of other wildlife species, especially songbirds and raptors. On this front it benefits from its location adjacent to Mexico. Many Neotropical species that visit the refuge cannot be seen in the United States except in southeastern Arizona, and this accident of geopolitical history structures bird-watching in significant ways.

AGUIRRE LAKE

Aguirre Lake, the grandest human creation of the Buenos Aires Ranch, appears to have been the original impetus to public use planning for the refuge. Ever since the days of Pedro Aguirre, the lake had been a destination for picnics, fishing, and swimming by area residents, and some owners had sold duck-hunting privileges to sportsmen. The masked bobwhite did not benefit from the lake, however, so the FWS did not initially pay it much attention. After shutting down its own experiments in 1981, the FWS hired biologists on contract to return to the ranch each summer to survey for surviving masked bobwhites. The 1982 survey focused only on the masked bobwhite, locating four pairs and five calling males (Ough and deVos 1982). Two years later the survey was expanded to include "all other birds encountered." G. Scott Mills (1984) found no masked bobwhites, but he identified 112 bird species, more than half of them (66 species) at Aguirre Lake, including 17 species associated with large bodies of water. (The heavy rains of late 1983 probably contributed to these numbers.) After acquiring the Buenos Aires, the FWS singled out these "species of special emphasis" as "an important asset" of the refuge. Intentional flooding was proposed for the refuge on the grounds that "these birds will be popular with the public and will thus contribute to the public use program in addition to the diversity goal of the refuge" (FWS 1986b: 9–10).

Refuge planners envisioned Aguirre Lake as the centerpiece of public use. Its location near headquarters meant that "a wide variety of habitats" were all contained within a small area at the core of the refuge. "The public will thus have the opportunity to view a great variety of bird species, including many rarely seen in arid lands such as egrets, rails, and plovers, as well as the elusive masked bobwhite. Mule deer are favorites with refuge visitors and these handsome animals will be regularly seen from the trails and tour route sited in this area. Coyotes, javalena [sic], and many other animals will frequent the area, to the delight of the human visitors" (FWS 1986b: 14–15). Converted to a visitors' center, the main ranch house would overlook this engineered bounty of habitats and wildlife. A seven-mile "interpreted auto tour route" would be created on old ranch roads, providing access to scenic overlooks and a gamut of interpretive signs (FWS 1986b: 15). Shortly after acquisition, the refuge installed small wooden signs to mark Aguirre Lake and the adjacent reservoir, which was dubbed "Grebe Pond."

Despite the damage done to headgates and dikes by the floods of 1983, Aguirre Lake continued to fill from heavy rains in the early years of the refuge, and it remained a major focus of public use efforts. Under the heading "Wildlife Diversity," the 1988 Annual Narrative Report announced that a male garganey, a Heermann's gull, eight Bonaparte's gulls, eight sandhill cranes, an oldsquaw, and two merlins had been sighted at the lake. The garganey "was the second record for this species in Arizona. Approximately 100 Arizona birders added it to their life list before it departed" (FWS 1988: 38). As the official list of birds sighted on the refuge has grown (to more than three hundred species today), a sublist has been kept for the lake.

Over the years, however, Aguirre Lake has virtually ceased to exist. Two of the three tributary watersheds are contained almost wholly within the refuge, and the increased mass of vegetation resulting from the lack of grazing has, by all appearances, kept rainfall from running off in quantities sufficient to fill the lake. A headgate on the third and largest tributary, Bailey Wash, was damaged in the 1983 floods and has not been repaired. In a sense, "original" conditions have been partially restored, but at the expense of Aguirre's most famous improvement and the water-dependent birds it once attracted. In the late 1980s, management attempted to pump groundwater to fill the lake, but this proved costly and ineffective. It also raised legal difficulties. The refuge held only *agricultural* water rights, descended from the days of the ranch. If these rights inadvertently created benefits for wildlife, the Arizona Department of Water Resources could not object. The refuge, however, had no legal right to pump groundwater expressly for conservation or wildlife objectives.

Since the floods of 1993, Aguirre Lake and Grebe Pond have both been dry, and the small wooden signs are now shrouded by young mesquites. A bench erected for fishermen prior to the refuge is likewise hidden, some twenty feet up a dike above the dry bed of Grebe Pond. The lake remains on most maps, however, and visitors occasionally show up with boats hoping to fish.

ANTELOPE DRIVE

The auto tour loop envisioned in the first planning process was never realized, perhaps because of a lack of funds. Instead a different route was designated in 1988, heading south from headquarters to the small border town of Sasabe.[2] The road was already in place, and contrary to plans, it was not improved with gravel; all that was required was the placement of signs indicating the route. The wildlife promoted by the route was a herd of eighty-six Chihuahuan pronghorns, which had been imported from Texas and released in the area the previous year. (The herd suffered large initial losses to predation; for several years aerial gunning and other techniques were employed to reduce the coyote population in the area during fawning season.) In a paradigmatic example of "staged authenticity," the refuge planted the sight (pronghorns) and erected markers (signs) to direct visitors to it (MacCannell 1976).

Apart from any issue of authenticity, the installation was biologically misleading in two respects. First, the Chihuahuan pronghorn is classified as a separate subspecies from the Sonoran pronghorn, and the Sonoran variety is more likely the one that occupied the Altar Valley in the nineteenth century. Sonoran pronghorns are too rare to allow transplantation, however. Subspecies level distinctions, on which the entire masked bobwhite program rests, were disregarded in this case. Second, pronghorns are not a type of antelope, but the tour route was named Antelope

Drive anyway, perhaps because most tourists do not recognize the distinction and are less familiar with the name "pronghorn." In any event, the refuge defined all use of the route as "wildlife observation," a bureaucratic category with official approval in the FWS. No interpretive signs were installed. The route has never become a major attraction.

ARIVACA CREEK AND ARIVACA CIÉNEGA

Three years into operation, fewer than one thousand visitors were registering at the headquarters visitors' center annually (FWS 1988). Refuge management attributed this discouraging record to inadequate funding. The 1989 Annual Narrative Report noted the "estimated population of 2 million people within three hour's [sic] traveling time from the refuge": that is, in Phoenix and Tucson. With an optimism reminiscent of nineteenth-century boosters, the report claimed, "Once appropriations become available for improving the roads and converting a large adobe residence into a visitor center, the refuge could attract an estimated 50,000 visitors annually. Similar natural attractions in and around Tucson attract upward of 800,000 visitors annually" (FWS 1989: 45).

Without water in Aguirre Lake, however, the refuge lacked the riparian habitat necessary to attract wildlife and people alike. It was at this time that the focus of public use efforts shifted to newly acquired areas of the refuge along Arivaca Creek. In the 1980s ecologists and environmentalists drew attention to the importance of riparian areas such as the ciénega and creek, where the presence of water in an otherwise arid landscape creates habitat disproportionately important to wildlife of all kinds (Hendrickson and Minckley 1984). The creek, however, sank underground several miles short of the Buenos Aires Refuge, so at first none of this high-value habitat was on the refuge. Almost all of it was deeded land and grazed by livestock. Shortly after the refuge was created, FWS realty officials developed plans to acquire the creek from the east boundary of the Buenos Aires all the way to Arivaca Lake, a man-made reservoir about five miles above the ciénega; these plans were later scaled back slightly.

Acquisitions began in 1989 with 225 acres of the ciénega; another 1,800 acres have been obtained since then. As with Brown Canyon a few years later (see below), these lands are neither masked bobwhite habitat nor Sonoran savanna grassland. The purchases were enabled by a peculiar bureaucratic logic. The regional office organizes potential acquisitions according to habitat types, with priority given to those that are underrepresented in the refuge system as a whole. The Buenos Aires Refuge is classified as a "desert grassland" refuge. Because this *type* is a high priority, Buenos Aires acquisitions are routinely pushed to the head of the list, whether or not the parcels acquired are actually desert grassland.

Most of the public use objectives earlier linked to Aguirre Lake were transferred to the ciénega and creek, especially after a full-time outdoor recreation planner (ORP) was hired to direct public use programs in 1993. By this time the initial opposition to the refuge's creation had subsided, though relations between area ranchers and refuge management remained poor. A few ranchers had sold parcels along the creek to the refuge, but many others viewed refuge expansion with suspicion (especially because acquisition plans had been drafted without consulting landowners, several of whom were outraged to discover that the FWS was already referring to their properties as future refuge lands). The ORP's work appears to have enflamed the smoldering coals of this feud. It was her job to develop trails, interpretive signs, and programs and to attract more visitors to the refuge generally; gaining media attention was an integral part of her strategy. In promoting public use at the ciénega and creek, she chose to stress the removal of livestock as restorative of "natural" conditions; like the masked bobwhite before, the wildlife at these places became symbols whose value derived from their putative opposition to cattle. "[T]he fall of 1993 marked the year the Buenos Aires was 'discovered', as several magazine, newspaper, and TV spots created interest in refuge wildlife and trails," wrote the ORP (FWS 1993: 53). This media coverage and the anticattle message it conveyed helped to provoke renewed criticism from cattlemen in the mid-1990s (see chapter 9).

The substitution of "nonconsumptive" visitors for "unnatural" livestock was most conspicuous along Arivaca Creek, where the refuge purchased 1,641 acres (470 deeded and the rest State Trust lands under lease) from Eva Antonia Wilbur-Cruce in 1990. Her family had homesteaded there in 1862, but the elderly Ms. Wilbur-Cruce had long since moved to Tucson, leaving behind a small herd of unusual horses. The

1990 Annual Narrative Report summarized the situation without a trace of irony:

> When the Wilbur-Cruce property was acquired, it was general knowledge that a herd of horses was native in the area. The Refuge took immediate action to remove the animals. Inspection by equine experts across the country determined that the horses were the last known herd of Spanish horses left in North America. The horses were originally brought to the Wilbur-Cruce property around 1881. They were purchased from a herd in Mexico which was descendants [sic] of Father Kino's horses. These horses were used to settle the Southern Arizona Territory at the close of the 18th century. (FWS 1990: 4)

"There are very few strains of horses worldwide that have been kept pure, and this herd represents one of them," noted a veterinarian and geneticist who inspected the herd. "These are the horses you see in 14th- and 15th-century Spanish paintings" (*Arizona Republic,* 30 April 1990). Area residents, Ms. Wilbur-Cruce, the American Minor Breeds Conservancy, and the Spanish Mustang Registry all appealed to the FWS to allow at least some of the horses to remain, pointing out that they could even become a tourist attraction.

Being rare, pure, and "native" (i.e., wild and surviving without human tending) was not enough to bring the horses within the protected enclave of wildlife, however. Indeed, their "wildness" was a strike against them: As horses, they were categorically defined as "domestic," and the fact that they had "reverted" to a wild state threatened the distinction on which the legal definition of wildlife rests. Regional Director Michael Spear ruled "that the refuge was purchased specifically for the re-establishment of the masked bobwhite quail, and . . . regulations prohibit introducing feral horses and burros on national wildlife refuges" (*Tucson Citizen,* 28 April 1990). FWS officials insisted that the horses were private property not included in the sale of the land; they were therefore not wildlife but trespass livestock. The refuge manager extended this logic when he said, "It's pretty obvious that lions are taking some of the colts; that's why we think it's important to get them off the refuge as quickly as possible"

(*Arizona Republic,* 4 June 1990).[3] In the end, the horses were donated by Wilbur-Cruce to minor-breeds groups and removed from the refuge. The herd, about one hundred animals, was too large for the pasture they had occupied, and after several dry years, the area was severely overgrazed. The ORP may not have known it, but she later used photos of the land taken in 1990 to represent prerefuge conditions generally, when in fact the Wilbur-Cruce property was nothing like most of the Buenos Aires at the time of acquisition by the refuge.

Public use developments at the Wilbur-Cruce property were fairly modest. Whatever damage the horses had done, the wildlife were there and no habitat manipulation was required for them, only for the human visitors who would come to see them. A parking area was cabled off and a trail was constructed leading down the creek, past the old Wilbur-Cruce home, and then up a nearby hill. It was named "Mustang Trail," as though to erase the conflict by a token memorial. The 1995 Annual Narrative Report described the trail this way:

> Vegetation is recovering from overgrazing and birds are abundant in the diverse vegetation. Overhead, cottonwoods hold grey hawk *[sic]* and other raptor nests. Woodpeckers and sapsuckers work the older dying branches for insects. In migration, colorful warblers flit among the thorny shrubs. Summer tanagers and yellow-breasted chat are two of the most colorful and noisy birds you can find in summer. Deer and mountain lion travel in and out of the area for water and a troop of coatimundi is sometimes seen in winter. (FWS 1995: 64)

An interpretive sign was placed at the trailhead in 1995. Alongside the road, another sign was erected to alert motorists to the site (fig. 18). The iconic binoculars shown on the sign are the symbol of the "Watchable Wildlife" campaign, launched by the U.S. Department of the Interior in the early 1990s in conjunction with the refuge slogan mentioned earlier. The sign is brown, the color of state-sanctioned cultural and recreational attractions. It is not about the law but about leisure or edification. If this were a private sign, we would read it as an advertisement, reasoning that because the sign cost money to erect, no private party would go to the expense without intending to recoup the cost, and then some, through its effects. We would expect to see the promised sight in exchange for money. This is a state sign, however, and its logic is slightly different. The cost of its installation was paid by a public entity, presumably to promote a public good. The attraction it promotes stands apart from the crass exchange of money, and although the sign creates an expectation, it makes no promises. The wildlife is "watchable" here, but it is only a potential sight. This is not a zoo but "nature," and wildlife species viewed in nature carry special charismatic value: Seeing them is a gift bestowed as though by grace on the viewer. This value increases both with the perceived rarity of the animal (mountain lions being higher in value than deer, for example) and with the perceived lack of artifice in the setting.

The sign appears to do nothing more than point to a naturally occurring bounty of wildlife. It does do more, however: It abstracts from actual sightings, collects them under the heading "wildlife," and effects a transfer of charismatic value to the refuge as a place distinguished from other places. Just as speed limit signs produce a space where certain laws apply, this sign officially authorizes a particular place *as* a place to view wildlife. In so doing, it makes the wildlife there into a public good. Unsighted, unrecorded wildlife species have no such existence. Through signs, "the state wields a genuinely *creative,* quasi-divine power" over space (Bourdieu 1998: 52, emphasis in original). Yet the greater the resources devoted to promoting the watchability of wildlife at a given place, the more important it becomes that sightings be perceived as "natural." Refuge officials suggest in brochures and presentations that wildlife is found there because it is a refuge, when in fact the refuge is there because of the wildlife. The habitat, as we have seen, has a long history of human impacts and manipulation. By implication, the sign suggests that wildlife species are not found on nonrefuge lands, when in fact the public observers are the ones who are lacking. To paraphrase Barthes (1957: 129), the principle of the sign is to transform history into nature by replacing human agency with atemporal essences. The Watchable Wildlife campaign aims to convert the charismatic value of viewed wildlife into symbolic capital—to make the FWS appear responsible for their presence—and thence into bureaucratic capital.

Figure 18. Sign directing motorists to Arivaca Creek wildlife-viewing area at the old Wilbur-Cruce property, 1997. (Photograph by the author)

Public use facilities were more ambitious upstream at Arivaca Ciénega, which has become the most popular public use site on the refuge. In 1993–94, summer Youth Conservation Corps crews constructed boardwalks across the marsh to complete a 1.5-mile loop trail. The trail guides visitors from the parking area around the ciénega, including a stop at an observation platform fitted with heavy-duty (i.e., vandal-proof) spotting scopes. The platform overlooks a small, man-made pond at one corner of the ciénega that is frequented by the kinds of water-dependent birds observed at Aguirre Lake in 1984. Planning for the boardwalk aspired to control the spatiotemporal experiences of visitors in detail:

A new trail plan was developed to correct several problems at the Ciénega, including: the existing trailhead was hard to find and sometimes cut off by water, falling branches from cottonwoods at the trailhead posed a safety hazard, we wished to reduce traffic near gray hawk nesting areas, the old trail came out next to an untidy residence, the old trail "loop" ended in town (¼ mile from trailhead), old loop was frequently under water, we wished to approach the pond with the sun at your back for morning birding. (FWS 1993: 54)

The new boardwalk succeeded in attracting media attention from a major Tucson newspaper (*Arizona Daily Star*, 28 November 1993), which refuge officials credited for a surge in visitation the following month. "We hadn't encouraged a lot of public use of the refuge in the past, because we wanted to put our emphasis on the masked bobwhite quail," the refuge manager told the reporter. "But now we have an outdoor recreation planner . . . and we're encouraging people to come out and enjoy the wildlife and natural environment." The ciénega "is slowly gaining a reputation as one of southeastern Arizona's 'destination birding sites,'" wrote the reporter, apparently quoting the ORP. This had been the FWS goal since 1985. The only drawback was the location of the ciénega more than twenty miles' drive from the visitors' center at headquarters, where facilities were (and are) still being developed as though Aguirre Lake were the main attraction. Most birders do not bother to drive the extra distance (FWS 1993: 53).

HUNTING

Hunters make up some 10 to 15 percent of public use on the refuge, according to refuge figures. I did not make hunting a focus of my research, and I limit my discussion here to a handful of remarks. First, the AGFD has jurisdiction over hunting in Arizona, including federal lands such as the refuge. It manages hunting through a system of "game management units," within which censuses are conducted and permit levels set. Three such units overlap portions of the Buenos Aires Refuge.

Second, hunting has been the subject of ongoing negotiations, sometimes collaborative and sometimes contentious, between the refuge and the AGFD. The result is a complex system of spatiotemporal regulations keyed to different game species. All quail hunting is forbidden on the refuge in order to protect the masked bobwhite, and areas surrounding houses and public use facilities are closed to all hunting for human safety.

Open seasons for the major game (white-tailed deer, mule deer, and javelina) are temporally staggered depending on the species and the type of weapon used (bow and arrow or firearm).

Third, although FWS policy officially recognizes hunting as a priority public use, the AGFD has long felt that refuge management downplays or neglects hunting relative to nonconsumptive public uses such as bird-watching. Disputes over the size and health of the mule deer herd, the maintenance of artificial waters, permit types and numbers, and hunter access to refuge roads and lands have recurred. Negotiations completed in 1993 resulted in the designation of two of the relevant game management units as "quality hunt areas," where mule deer permits have been reduced to decrease hunter density and raise success rates; the Buenos Aires Refuge is now viewed as one of the best deer-hunting locales in Arizona (FWS 1998). Refuge management would like to see the refuge made into its own game management unit, but the AGFD has refused.

The opposition between hunting and bird-watching may be understood in two ways. Most obviously, one involves killing and the other does not. Hunters take the actual bodies of animals, converting them from public property into private property, whereas bird-watchers take them only symbolically through visual encounter. An animal can be seen any number of times, but it can be killed only once. Aside from this, however, the practices of hunting and birding have a great deal in common: the careful search for wild animals, the perceived closeness to nature, and the ritualized practices repeated over time among small groups of friends or relatives (Nelson 1997). In fact, studies in the 1970s found that hunters value these "nonconsumptive uses" of wildlife more highly than the meat they procure (Gray 1993: 97). Moreover, as we will see, birders also appropriate their quarry as a kind of private property.

Because hunters and bird-watchers rarely seek the same animal species, they ought in theory to be able to coexist, but a social opposition intervenes: Hunters and birders are by and large distinct groups of people with different dispositions regarding proper relations to wildlife. Killing animals for food highlights the carnivorous and predatory—some might say the animal—aspects of human existence, whereas the visual appropriation of wildlife, particularly birds, partakes of a more aesthetic, contemplative ethos. Viewed in terms of the opposition between consumptive and nonconsumptive activities, hunters fall into the same category as ranchers, loggers, and miners: Their actions appear motivated by appetite rather than refinement, by need rather than election. Bird-watchers, in my experience, tend to assimilate hunting to working class values of physicality, virility, violence, and crudity.

COUNTING VISITORS AND WILDLIFE

As with every official activity on the refuge, public use must assume a paper form for purposes of bureaucratic mediation. Like cattle on the Buenos Aires Ranch, visitors are measured in standardized units: Visitor-days have replaced animal-unit months. Although most wildlife species are not subjected to the intensive censusing employed for the masked bobwhite, they too are documented. Precise measurement is difficult in both cases because neither people nor wildlife species reliably self-record their presence. In the early 1990s, in conjunction with the Watchable Wildlife campaign and in response to congressional pressure to justify its funding requests, the FWS increased its efforts to document public use. The results clearly indicate a decision to emphasize nonconsumptive activities such as birding (FWS 1997; see below). The Buenos Aires Refuge is a case in point.

Wildlife species are documented and counted within the framework of Linnaean taxonomy (see chapter 1), chiefly during biological research conducted by scientists. Threatened, endangered, and rare species receive the highest priority, both for research projects and for ongoing monitoring purposes. Sightings of species never found on the refuge before or known as extremely rare require a greater level of documentation or authority in the viewer. When a scientist traps a new species, photographs it, and writes down details of the sighting, it is promptly added to the refuge list and remains there indefinitely. In contrast, an amateur who catches a passing glance of an animal at a distance, as has occurred with jaguars several times, will be doubted and asked for details to substantiate the claim. The greater the documentation, the more credible the sighting, and only one credible sighting is needed to place a species on the official list.

As evidenced in the Annual Narrative Reports, the arrival of the ORP in 1993 resulted in a heightened emphasis on species lists, especially for birds. A brochure she developed lists bird species by family, in accordance with the American Ornithological Union's classificatory system. It further distinguishes each species according to the location on the refuge where it has been sighted and the temporal parameters of its presence. Symbols indicate cases of rare or undocumented sightings and species known to breed on the refuge. This representational compendium is a standard item for use by bird-watchers and an important index of the refuge's relative significance among "destination birding sites." The total number of species sighted can only grow because none are removed with the passage of time; it is prominently featured on the front of the brochure. The abundance of each species is of less importance in most cases. Only scientific researchers trained in censusing techniques and statistics can produce credible figures on abundance, and only selected species warrant such work. For birds the primary source of population figures is the annual National Audubon Christmas Bird Count, in which skilled birders report all the birds they see during a day of intensive, coordinated searching.

Counting visitors to the refuge is almost as gross in its margins of error as counting masked bobwhites. Only at Brown Canyon, where access is controlled, can accurate numbers be compiled; elsewhere estimates must be generated from limited data. Registers are kept at the headquarters visitors' center and at a small office in Arivaca, but there is no way to compel visitors to sign in, and many do not enter the offices at all. In 1995 the ORP decided to double the register figures to account for those who do not sign in, causing the official figure to leap to seven thousand visitors, compared with three thousand the year before (FWS 1995: 61). At the ciénega trail and on the driveway to headquarters, hidden electronic counters were installed in 1994 to count passing walkers and vehicles, respectively. (These numbers are adjusted to account for multiple passings by single visitors and for the number of passengers in each vehicle.) Hunter use is calculated as a fixed percentage of the overall number of permits issued for the three game management units that overlap the refuge, although it is impossible to know if that percentage accurately reflects reality, and permit numbers are controlled only for certain game species.

As with the masked bobwhite, visitor numbers must nevertheless be generated and presented as if accurate counting were possible. It is in the interest of the refuge, particularly the public use program, to demonstrate maximal public use. Visitation numbers increased sharply during the first few years following the arrival of the ORP: up 30 percent from 1994 to 1995, when 25,500 visitors were counted (FWS 1995: 61). How much of this rise was due to actual increases in visitation and how much was due to improved or altered counting techniques are impossible to say. In some contexts, a further adjustment is made to achieve at least the appearance of greater objectivity. What, after all, constitutes a "visit"? Some temporal measure must be imposed to account for different lengths of stay. A visitor-day on refuges is calculated as eight hours long (FWS 1997). Different categories of use are adjusted according to estimates of how long each activity typically takes. It appears that this adjustment, which would increase the relative significance of hunters, who typically stay overnight, has not been made in the official figures for the Buenos Aires Refuge.

BROWN CANYON

Nowhere have the contradictions of the Buenos Aires public use program been more conspicuous than in Brown Canyon, a steep and rugged notch in the east slope of the Baboquivari Mountains. Named after Rollin Brown, who grazed livestock and homesteaded near the mouth of the canyon in the 1880s (and who should not be confused with Herbert Brown), the canyon rises quickly through a series of vegetation types dominated in turn by mesquites, sycamores, hackberries, several species of oaks, and junipers. At a few locations geological formations push water to the surface, but most of the time stream flow is limited, and consequently, so have been the canyon's human uses. Only four houses and a scattering of outbuildings stood in the canyon when the FWS launched acquisition efforts in 1993; three of the four now belong to the refuge.

Near the canyon mouth lived a wildlife artist renowned for his exquisitely detailed portraits of animals painted entirely from memory. He also painted Edenic landscapes with improbable assemblages of wildlife arrayed around pools or under big trees. Brown Canyon afforded him views of mountain lions, white-tailed deer, bobcats, and all manner of raptors and songbirds. His vigilance in defending his private land against trespass by hunters—he reportedly sat at his gate on weekends in season and rebuffed all comers—contributed to a myth among hunters that deer in the canyon are abundant and tame. His elegant home is now slated to become a visitors' center. Farther up-canyon is a large, two-story, stone-and-block building, whose history includes use as a hunting lodge, a secluded home for the large family of a popular musician, a cooperatively owned vacation retreat, and, according to legend, a brief stint as a brothel. As part of the refuge, this building has become the Brown Canyon Environmental Education Center, a.k.a. "the lodge."

Above the lodge a small ranch outfit was last owned by a commodities broker who moved to Tucson from New Jersey in the early 1960s. He named it the Happy Buzzard Ranch and used it as a weekend getaway and tax shelter. The steep canyon walls kept his cattle from roaming, and he tended them so irregularly that they eventually became feral. Perhaps a dozen of their descendants still roam the upper canyon, withstanding the mountain lions and repeated human efforts to capture them. The refuge has refrained from shooting them, even though they are potentially a menace to humans, for fear of provoking a backlash from cattlemen, who bear no allegiance whatsoever to the former owner. The most striking legacy of the Happy Buzzard Ranch, however, is the mile or so of dry stone walls constructed by undocumented Mexican laborers. Ranging from two to three feet thick and four to six feet high, the walls line the road, enclose the ranch house and a well, support old gravity-fed waterlines, and in some places serve only ornamental purposes. Apparently the "rancher" had them built in order to keep the Mexicans busy. Refuge officials, whose inclination is to remove all traces of human occupation other than the visitors' center and lodge, concede that the walls will have to stay because removing them would result in more damage than benefit.

Refuge acquisition of the canyon occurred rapidly, spurred on by the supposed threat of private development and enabled by the refuge's priority status in regional land acquisitions. The wildlife artist approached the refuge in late 1992 and expressed an interest in selling his property; for reasons of health, he and his wife needed to move somewhere with phone lines, electricity, and better access to the outside world. The FWS responded that it would pursue acquisition only of the entire canyon, not wanting the management difficulties of coexisting with other owners. Plans went forward when all but twenty-five acres were offered for sale

(FWS 1993). The refuge manager reports that regional realty officials instructed him not to buy the buildings because they would become maintenance burdens for which staffing and funding were inadequate, but the owners refused this arrangement, and the FWS pushed ahead anyway. The ecological justifications for purchase resembled those for Arivaca Creek: As a riparian area, the canyon provides important habitat for wildlife, and it adds to the "diversity" of the refuge as a whole. Moreover, it creates a corridor of refuge land from the Baboquivari Mountains on the west to the ciénega on the east, although this argument presumes that wildlife prefer to migrate on refuge lands.

Everything suggests that public use was the overriding motivation for acquiring Brown Canyon. In 1995, with acquisitions still ongoing, the Annual Narrative Report announced, "It is expected that the canyon will be a major attraction with its excellent birding and scenic beauty. Planning will consider sensitive development as a premier watchable wildlife area" (FWS 1995: 64). The regional office's warnings have proved prescient, however. The very features that inhibited private development in the past—poor access, no on-the-grid electricity, no phones, and limited water—now pose barriers to refuge ambitions. It was relatively easy to secure funding for acquisition, but no increase in the refuge's operations budget accompanied the new land and facilities. The lodge has required a new septic system, solar power system, backup generator, roof, water tank, appliances, and furnishings as well as extensive remodeling. These expenses, largely unanticipated, have drawn resources away from other refuge programs. While the head biologist writes grants to obtain funding for research, for example, more than seventy-five thousand dollars was spent improving public use facilities in Brown Canyon during fiscal year 1997 alone. Communications have been a chronic problem, and the unimproved road up the canyon is deteriorating under increased use. The refuge has no money to staff the facilities, so it must rely on volunteers who exchange their labor for free housing.

The uniqueness of the canyon and its draw for visitors rest almost wholly on its relative lack of human use or presence, but it has seen far more humans under refuge ownership than ever before. The sole remaining private residents report that wildlife is already less common in the main part of the canyon than before refuge ownership, even though public access has thus far been on a reservation-only basis. There is some evidence that bird feeders placed at the lodge and visitors' center for the benefit of birders may be attracting cowbirds and thereby impacting nesting success. As with the refuge in general, no baseline ecological data were collected after acquisition, so rigorous evaluation of refuge-related impacts is impossible.

Although a new addition to the refuge, Brown Canyon is symptomatic of problems afflicting the refuge system as a whole. Some $575 million in construction and improvement projects are backed up in the system, according to the U.S. Department of the Interior. The solution being implemented for the canyon is also representative of a larger trend. With a single access point, the canyon is the only place on the refuge where visitors can easily be compelled to pay out of pocket for their experience, and the refuge nominated it for inclusion in the U.S. Department of the Interior's experimental fee-area program created by federal law in 1996. Under the law, 80 percent of fees remain on the field station where they are collected, giving managers a powerful incentive to develop and maximize public use. Lawmakers reasoned that visitors derive a "greater benefit" than other taxpayers and that they should accordingly bear a larger share of the cost. Watchable wildlife has thus become a commodity in the full sense of the term, produced by refuges and other public lands and consumed by a free market of visitors. We must now consider what motivates these visitors.

BIRDING

As a government agency, the refuge is officially secular. The Brown Canyon Environmental Education Center, for example, is by policy available only to groups engaged in environmental education and not for such things as Bible study or church retreats. The presence and meaning of wildlife on the refuge are officially interpreted in rational-scientific terms structured by Linnaean taxonomy. Visitors who come to see wildlife are understood—and usually understand themselves—to be pursuing a form of recreation rather than spiritual enlightenment. Among the visitors I encountered in thirteen months in Brown Canyon, most explained their activities in terms of pleasure: in being outdoors or among friends, in the challenge of seeing and identifying new or beautiful species, or in getting away temporarily from their ordinary lives. Yet wildlife viewing, like tourism generally (MacCannell 1976), is a form of secular ritual, one that apprehends wildlife, and by extension the refuge, as embodiments or expressions of a sacred Nature.

The predominant type of wildlife viewing on the refuge is bird-watching. Birding (as it is more frequently termed among practitioners) has in recent decades become a hugely popular recreational activity (Gray 1993: 98), partly because of its flexibility. It can be pursued almost anywhere and at any level of dedication or skill. It does not require athletic prowess or great physical exertion, making it particularly popular among older age groups. The most fanatical birders openly confess a passionate or obsessive relation to birds; others take a more detached or passing interest. Few would deny, I think, the ritual aspects of their chosen avocation. I do not claim here to describe all birders, though I do believe that those I met on the refuge may be taken as representative for the purposes at hand. Most came from educated, middle- or upper-middle-class backgrounds; many were retired; virtually all were white.[4] Many described having provided feeders, special vegetation, birdhouses, or birdbaths in their yards at home to attract birds. They took weekend or holiday trips to places such as the Arivaca Ciénega or Brown Canyon, hoping to see a more or less specified set of species.

To attach symbolic or spiritual meaning to elements of the natural world—animals of all kinds, mountains, rivers, trees, etc.—and to elaborate rules for ritual and everyday conduct in terms of such "totems" are by no means unusual. The peculiarity of birding is that overtly spiritual or religious aspects are denied, even while its practices are codified and ritualized in unique and unprecedented ways. The overarching system that classifies bird life is understood in scientific rather than mystical terms, grounded in empirical biological research rather than divinity. Yet the value attached to birding as a recreational activity or avocation far outstrips science: Birding is not understood as research; although it derives historically from the endeavor to collect and classify species, it is not motivated by the pursuit of scientific knowledge. Birding instead produces pleasure or gratification through a particular way of approaching, understanding, and relating to the natural world.

The fundamental event of birding, as with other wildlife viewing, is the sighting, the moment of visual contact between animal and person. It is not simply seeing the bird but seeing it in a particular way, recognizing and identifying it as being of a certain (sub)species. The criteria employed to distinguish species are both visual and extravisual: size, shape, color, markings, behavior, location, and song or call. Something more than mere knowledge of the bird's identity is involved, however. Scientists conducting censuses or other research will accept a vocalization, a feather, or a nest as proof of a bird's presence, but in birding these criteria are only corroborative and do not constitute a sighting. The living bird must be seen. It is not a question of authenticity, as though only visual confirmation were real. It is rather as though a sighting were an exchange consummated by visual connection. Usually this takes some moments: A flutter in a tree catches one's eye, drawing a concerted look; binoculars magnify the power of one's gaze; the bird is obscured one moment and then finally is observed in full. Its appearance is received as a kind of gift of fortune or Nature; the birder reciprocates with recognition and appreciation. For beginners, a moment may pass before the exchange is completed; time is spent flipping through a field guide to confirm the bird's identity. For experts, even the briefest glance may complete the sighting. However one defines the value of wildlife viewing, it is produced in this exchange.

Between the beginner and the expert is experience, education attained through discipline of mind and body alike. In books, birders study the appearance, range, migration patterns, habitat preferences, and behaviors of different species; tapes and compact disks familiarize them with songs and calls. Most carry one or more field guides with them while birding. From various sources they learn where to go and at what time of day or year to see the birds they seek. Successful birding, however, requires careful control of one's body above all else. A quiet, slow pace is kept; voices are hushed as though in reverence. Good birders see things others miss; they raise their binoculars to their eyes without changing the direction of their gaze, ensuring that time is not wasted scanning for the bird again. They do not need to consult the books as often, and when they do they know where to look to confirm their sightings more quickly. Body, mind, hands, binoculars, and eyes work together with a kind of athletic coordination, manipulating sight with the unconscious grace of practice.

Most birders learn in groups by mimesis; only professionals or more ascetic birders regularly bird alone. Many insist that the companionship and group dynamic are among their favorite aspects of birding. In the group, rules of comportment are tacitly inculcated and enforced, usually through the establishment of a hierarchy of skill and authority. If the hierarchy is not already known, it quickly emerges as the better birders sight birds that others miss or do not recognize. The charismatic value of the birds becomes the charismatic authority of the skilled birder, which is constantly tested and reproduced. Only one person in a group makes the generic vocalization *psshh psshh*—known as "pishing"—which is intended to elicit responses from birds and thereby disclose their locations; this prerogative is typically granted, without explicit discussion or assignment, to the individual deemed the best birder. In turn, the implicit hierarchy of birders structures the identification of birds sighted. Each individual gets a slightly different look at each bird or no look at all. Because it is understood that the bird must be of one and only one species, a process of consensus formation and sometimes debate ensues:

A: "Are those Crissals?" B: "I didn't get a good look at them." A: "I know they're thrashers, but are they Crissals?" (Pauses and gazes through binoculars again.) "There are three of them . . . no, two. They're not curve-bills." (Calls in C to look. Debate ensues. Finally C admits she was looking at the wrong birds: towhees. C declares the birds to be curve-bills, and A and B assent.)

It is, of course, possible that all may be wrong, and in cases where the sighting carries larger significance, it is the authority of the most highly regarded birder in the group that determines whether others accept the sighting as credible. In most cases, though, the group's decision is the final judgment of what the bird is. Books carry substantial authority in arbitrating identification, but highly reputed birders may occasionally overrule a book by pointing out a flawed depiction or invoking further knowledge (e.g., of plumage variations). Ideally, the hierarchy works itself out to everyone's satisfaction: the best birders help the others to achieve more and better sightings and, in return, earn their respect and admiration. Often a less skilled birder will catch the first glimpse of a bird, and a more experienced one will aid in consummating the sighting, to the gratification of both. I have seen cases, however, where two individuals entered into rivalry for leadership, and in one case the group split in two, each with its own leader.

A complex hierarchy of birds likewise emerges and is reproduced through the collective efforts of birders, with rarity as the fundamental measure of value in a sighting. The most highly valued sightings are of the birds that are hardest to sight, meaning that good birds and good birders go together. The 1993 Annual Narrative Report announced that "quite a few 'good birds' like the green kingfisher, least tern, white-rumped sandpiper, and painted redstart" had been sighted at the ciénega that summer. "The Tucson birding hotline attracted good birders who spread the word. Until the birding 'bibles' of the area . . . are rewritten to include the refuge, word of mouth will be our best PR" (FWS 1993: 59). It is not the rarity of the bird but the rarity of the sighting that matters. As in taxonomy, geopolitical boundaries establish a spatial frame for classifying sightings; the value of any given sighting is assessed relative to the field of all sightings within this spatial framework. When a black-capped gnatcatcher was sighted in Brown Canyon, for instance, it set off a flurry of visits from serious birders, not because it is a rare bird—it is abundant in Mexico—but because it is rarely sighted in Arizona. "Political boundaries are very important to avid birders," as one ornithologist told me. "They keep a separate list for each state." The same framework extends into law as well: A handful of listed endangered species (the Sonoran chub and the cactus ferruginous pygmy-owl, for example) are abundant in Mexico. It is this socially constructed definition of rarity that makes southeastern Arizona into a "birding hotspot" because many Neotropical species are found nowhere else in the United States.

Beauty and nativeness are secondary measures of value among sightings. Even more than rarity, these judgments are socially constructed and often carry considerable social and historical baggage. Valued birds such as the elegant trogon, warblers, or the varied bunting suggest that beauty is largely a function of brilliant or extraordinary plumage; it can also correspond with special physiological attributes, as in the case of hummingbirds. Although birders do not usually call any birds "ugly," carrion eaters such as vultures and ravens often carry strong negative aesthetic judgments. Raptors, once reviled for killing innocent song-birds, have assumed new value following the collapse (and subsequent rebound) of many populations from DDT in the post–World War II period.

The value of nativeness derives largely from its opposite: birds that are "non-native" or "invasive." The negative value of these birds stems from biological relations among different bird species. Rock doves (domestic pigeons) and European starlings, for example, are "trash birds," both because they are commonplace and because they accompanied the invasion of Anglos across North America and the consequent creation of "artificial" habitats for them. The arch invader in southern Arizona is the cowbird, named for its symbiotic relation with cattle. It evolved to migrate with the bison herds. Rather than build its own nests, it lays eggs in the nests of other birds; its chicks subsequently out-compete the other young for food. The cowbird is thus simultaneously a symbol of cattle and a parasite upon more highly valued songbirds. Both beauty and native-ness tend to reinforce judgments of rarity; the surest sign of a "trash bird" is that it is easy to sight by virtue of its habits or numbers.

As mentioned already, some birders keep lists of their sightings subdivided by state and country; in this way they accumulate the value of individual sightings into a numerical measure of their birding accomplishments. Life lists are the private equivalent of the species lists promulgated by the refuge. "Listing" is something of a controversial issue among birders because "listers" are seen by some as bearing an improperly competitive attitude toward birding, akin to measuring the value of a library by the number of books it contains or the value of a person by his or her wealth. The debate raises the question of the social meaning of birding. Is it purely about the appreciation of nature, or is it a competitive sport? Is it about birds or about people? The answer can only be both. Some birders do seem to motivate themselves by comparing their sightings with those of others; almost all birders evaluate their own trips, one to the next, by the quality and quantity of sightings obtained. Local chapters of the Audubon Society routinely capitalize on the competitive potential of birding through "bird-a-thons," fund-raisers in which participants sign up sponsors who donate a certain sum of money for each species sighted in a twenty-four-hour period. In the Audubon-sponsored Christmas Bird Count, the verb "to get" was frequently used with a bird species as the direct object, as in "It would be good to get a gray flycatcher." Sightings become a form of symbolic, private appropriation.

Whether or not one keeps a life list, birding is fundamentally structured by the official scientific classification of bird life (presently some ninety-seven hundred species), and even birders who criticize listing display an astonishing memory for their personal sightings. For all dedicated birders, sighting a species for the first time is a memorable event, and the aggregate memory of these events becomes a complex spatiotemporal structure of their personal relation to the larger natural world. The higher the value of the sighting, the more detailed are the memories of the trip as a whole: when, where, with whom, the weather, the other sightings that occurred during the trip, etc. Each sighting occurs between a particular viewer and a particular bird at a particular time and place, yet it rises seamlessly to the level of the universal system of taxonomy, becoming a timeless, impartial apprehension of Nature legitimated by the secular truth-value of science. Every birder's sightings are at once individually meaningful and universally comparable. For biology, the abstraction inherent in viewing individual specimens as tokens of essentialized types is a methodological step; transposed to the world of birding, it provides an ontological bridge between contingent individual experiences and an abstract Nature. One is not merely seeing a bird but coming closer to the nature of all birds, and by extension, to Nature itself.

Unpredictability is a necessary condition for the charismatic value of wildlife; this is especially true for birds, whose mobility and long migrations make for tremendous variability over space and time. No two trips are the same, even if one returns to the same location many times, and there is always the chance of a new or rare sighting:

A: "Someone said there are loons at Parker Canyon Lake." B: "I went to Nogales Lake to get the hooded mergansers. I saw the mergansers one week and went back to see them again. They were gone, but there was a wood duck instead. I'll take a wood duck in place of a merganser any day: they're so beautiful I could look at them all day."

If sightings were guaranteed, their quality as gifts would disappear. A professional birding guide relates a saying commonly used when a trip has not yielded an expected sighting: "If we'd known for certain that we'd see it, we wouldn't have made the trip." Every trip is a leap of faith, so to speak, an occasion for testing one's fortune, and this is what gives birding its enduring appeal, one that can become addictive in ways rather like gambling.

In its combination of unpredictability, particularity, and universalism, birding bears a striking affinity with Protestantism. Individual grace inheres in personal accomplishments, realized in the face of uncertainty and measured against a universal system of values. Just as the tenet of predestination motivates the Protestant, paradoxically, to perform good deeds, the birder subscribes simultaneously to the charismatic unpredictability of the sighting and the devotional exercises of cultivated skill. A good sighting is both a revelation afforded by the bird and a reward for one's individual efforts. Nature, like God, is ultimately in control, but one can maximize the likelihood of grace by self-discipline, diligent practice, rising early, pious study of books (recall the allusion to "birding bibles" cited above), and a humble immersion in the world. Opportunities for grace are everywhere if one only learns to grasp them. It is as though religious principles and precepts have been transposed to an officially secularized realm of Nature. Nothing illustrates the parallel more clearly than the debate over listing, with its duality of pious virtue and invidious virtuosity.

The larger spatiotemporal framework of birding trips further reinforces their ritual significance. Trips resemble nothing so much as pilgrimages from the profane everyday world to sacred spaces where purity resides and revelations are found. For visitors to Brown Canyon this element was particularly pronounced. Access was restricted, and people could visit only as members of a group with reservations made ahead of time. By car they traveled the fifty miles from Tucson, arriving at a locked gate. There they would consolidate into a few vehicles, leaving the rest behind to minimize impact on the canyon, as if to protect its purity. As they ascended into the canyon, crossing the creek several times, the road became more rough and the sense of being remote, beyond the reach of pavement, telephones, and shopping centers, became more acute. An immense mesquite tree next to the road—one of the largest in Arizona, according to refuge officials—suggested antiquity and a lack of human impact.

Upon arriving at the lodge, with a view down canyon to the Altar Valley and upward to the monolithic Baboquivari Peak, home of the Tohono O'odham (Papago) creator spirit I'itoi, the visitors' first words were almost invariably about the canyon: "What a great place! What a beautiful place!" From the lodge, the pilgrimage continued on foot, ever higher into ever more pure Nature. The old road gave way to a footpath, which arrived finally at the arch, a natural rock bridge some forty feet high that spans the creek bed like a flying buttress. No matter what the particular interest of the group—birds, snakes, photography, butterflies, plants, or simple recreational pleasure—almost all took this walk to the arch, then turned around and returned to the lodge. It was as though some specific destination were necessary to give the trip a thematic climax, and this piece of natural architecture was a kind of shrine.

Back at the lodge, birders often sat on the porch comparing notes of their sightings. On one occasion, an experienced birder confessed that she had left cowbirds off of her list for the day. They were conspicuous, but she preferred not to grant them the status of having been recorded. Upon leaving, the visitors' departing words almost always stressed time— "We had a great time. What a wonderful visit!"—fusing together the spatial and temporal aspects of the journey.

As a pilgrimage, visiting the canyon derived its value by opposition to the ordinary world. References to "the money chase," long hours in the office, traffic, polluted air, and the artificiality of urban life were commonplace in conversation, and many visitors were curious to know what it was like to live in this idyllic antiworld. They invariably came to the canyon in the mode of leisure, and many seemed to envision life there as a perpetual weekend. Viewing the canyon through the lens of their own experiences, they transmuted the temporal dichotomy between work and play into a spatial one between home and canyon, city and refuge, and thence into a social one between congested humanity and unpeopled Nature. The vast majority of those who expressed an opinion believed that Brown Canyon should remain locked up. Some explained their position by referring to Madera Canyon (a major birding destination in the

Santa Rita Mountains south of Tucson), which they feel has become overrun with people who disrupt the birds by making too much noise or letting their dogs or children run freely. They expressed an almost cynical apprehension of market society, even where it extended into their chosen activity:

> A: "It's a spotted [towhee] in Arizona, a rufous-sided in California."
> B: "Oh, great, so now we have different names for different states!"
> C: "They have to change the names so we have to keep buying guide-books."

The value of bird sightings can be understood only against the larger backdrop of contemporary society. There is no disputing that birding is a leisure activity and that it fulfills a need to get away from or compensate for life in the everyday urban world. But how does birding, in particular, fulfill this role? It is easy to argue from the foregoing analysis that birding is a highly commodified social practice. It fetishizes species, both by isolating them as the fundamental units of all bird life and by assigning each one a relative value, which birders appropriate as a kind of private or personal possession. It elevates rare sightings much as the market for cultural distinction elevates artworks or fine automobiles. Especially for "listers," the accumulation of sightings is as limitless a challenge as the accumulation of wealth. Very wealthy birders are now so numerous that they support a small industry of professionally guided, organized birding tours to exotic and remote places around the world. One such guide tells of customers offering several thousand dollars to be led to particular birds in Arizona; one woman requested a private trip with the objective of seeing a five-striped sparrow as her six hundredth bird. It is no secret that birding is among the more refined bourgeois pastimes.

The symbolic values associated with birds run against the grain of commodification, however. Birds are "free," both in a quasi-political sense ("I'm as free as a bird") and in a monetary sense because they are a public good. Money spent to see them is never regarded as a "price" for each bird but rather as a secondary (and insufficient) measure of the "value" or "worth" of a sighting. A professional guide relates that some of his customers have mused, "This trip cost me twenty-five hundred dollars, and I went to see three particular birds; therefore each is worth about eight hundred dollars." Above all, bird-watching is understood as leisure, defined in opposition to work. To go birding is to retreat from the press of moneymaking and commodity exchange. The more freely one can afford to spend money to go birding, the more important charismatic qualities become: embodied skills, elaborate but unspoken codes of conduct and demeanor, and a refined aloofness from vulgar topics of business or money. A volunteer who staffed the refuge visitors' center in Arivaca remarked that birding allows people to talk endlessly with each other without encountering conflicts: "It can be a yellow cowbird or a red cowbird, but that's all they can disagree about."

Many birders feel that bird-watching is an important educational activity for advancing the causes of environmentalism. Birding heightens one's appreciation of nature and one's sensitivity to environmental issues, they say. It is true that birds are an important and ubiquitous indicator of the health and functioning of complex biological systems, as Rachel Carson's *Silent Spring* persuasively demonstrated. But this hardly seems sufficient to explain the social dimensions of birding as a recreational activity. There are many other ecological phenomena that could be monitored for environmental quality—insects or soils, for instance— but none has captured the imagination of so many people as birding. Its popularity rests not on abstract political or environmental principles but on its qualities as an activity and social practice.

Whether or not it is understood in terms of ritual, the value of birding is realized as an inverted form of commodity exchange. Its contents are understood in opposition to commodities, its practices in opposition to work. These oppositions presuppose formal homologies between the logic of value accumulation in birding and the fields of social practice against which it is defined. If sightings were guaranteed, their charismatic value would collapse, the experience being reduced to a kind of purchase, as in the price of admission to a zoo. If weekdays were not defined as the time of work, weekends could not be the time of leisure when one "recreates." If one could not accumulate sightings differentially valued according to rarity, they could not stand in opposition to commodities produced, exchanged, and accumulated in and through the capitalist economy. Like the mediation of commodity value, the biological processes that determine the appearance of birds occur beyond our view; a chance, high-value sighting is the symbolic equivalent of winning the lottery or stumbling upon a masterpiece at a garage sale. "We have loaded upon nature, often without knowing it, in our science as in our poetry, much of the alternative desire for value to that implied by money" (Harvey 1996: 163). Birding purifies one's secular life by apprehending products of nature as anticommodities, a form of wealth that enriches "the human spirit" (Gray 1993: 106) rather than the private pocketbook. If it were understood as a sectarian religion, it would lose the abstract, quasi-scientific universalism on which its efficacy depends.

WILDLIFE AS BUREAUCRATIC CAPITAL

Converting wildlife into bureaucratic capital requires that the sacred aspect of sightings find expression in a form suited to secular, bureaucratic mediation. Counting visitors is one way to do this, but it appears that a further conversion has been necessary in recent years. In 1997 the FWS published a study that it had commissioned from two economists, entitled *Banking on Nature: The Economic Benefits to Local Communities of National Wildlife Refuge Visitation* (FWS 1997). The study attempted to capture more than just the direct economic impacts of refuges: "National wildlife refuges enrich people in a great variety of ways. Some benefits are relatively easy to quantify—to attach a value to—and some are not. How much does that young couple value their beachfront sunrise? Or the duck hunters their excitement? Can a dollar figure—a price tag, if you will—be attached to people's dawning understanding of the marvelous workings of the natural world?" (FWS 1997: i). To answer the last question in the affirmative, the economists asserted:

> There are two components to the value of any commodity—what you pay for it and the additional benefit you derive from it over and above what you paid for it. If there were no additional benefit, you would not buy it since you could spend your money on an alternative good that would give some additional benefit. Surveys of the general population bear this out: Almost always, respondents are willing to pay more than they are currently paying for recreational

opportunities. Economists call the additional benefit "Consumer Surplus." (FWS 1997: 3)

Of course, the "consumer surplus" is an entirely imaginary figure, but the economists nevertheless persisted in giving it quantified expression. Visitors to wildlife refuges were asked what they had actually spent on their trips and if they would have made their trips had the cost been "X dollars more than the amount you just reported? X was a different random amount for different respondents" (ibid.: 13). Answers to the first question were used to determine actual spending; the second question yielded an average consumer surplus per day for different activities (birding, fishing, hunting, etc.), which was then multiplied by the visitation figures for selected representative refuges. By adding together the consumer surplus and actual spending, the authors of *Banking on Nature* arrived at the "net economic value" of refuges.

It need hardly be pointed out that this method of determining consumer surplus biases heavily in favor of wealthier tourists. Because they can afford to pay more, they are most likely to report that they would have paid more for their experiences. The high income of birders—three times the national average (Kerlinger 1993)—and their tendency to consider sightings "priceless" only exacerbate this distortion. The economists neglected to mention this, noting only that regional variations occur: "Consumer surplus for fishing in California for example is $132 per day while in Iowa it is only $6 per day" (ibid.: 72). As it turned out, the consumer surplus for nonconsumptive public uses such as bird-watching ($245 million) dwarfed those of hunting ($38 million) and fishing ($89 million; ibid.). In the analysis of particular refuges, the sleight-of-hand employed in deriving these figures was elided. Under the label "net economic value," consumer surpluses were simply added to actual spending (refuge budgets plus visitor spending) to yield "economic effects per $1 budget expenditures"—the fundamental concept for a Congress dominated by cost-benefit thinking. In several cases, as much as half of these "economic effects" were entirely ideal: dollars people said they would have been willing to spend but didn't actually spend. In this way, the sacred value of wildlife, especially among well-to-do Americans, became bureaucratic capital for the FWS.

In a final irony, *Banking on Nature* propounded an economic model that explicitly denies that local economies can grow from within: All economic growth, the authors asserted, derives from "importing some income from outside the region" (ibid.: 4). "The economic benefits to local communities of national wildlife refuge visitation" thus reduce, in large measure, to a kind of idealist Reaganomics: attracting outsiders (the wealthier the better) to experience their imaginary "consumer surplus" in one place rather than another, as though it might "trickle down" to ordinary working citizens.

CONCLUSION

Acquisition of the Buenos Aires Ranch was as much as the masked bobwhite by itself could yield for the FWS. Further bureaucratic capital could be generated only by expanding the refuge's public visibility as a tourist destination, and the masked bobwhite was poorly suited to this purpose. Without perennial water, moreover, even the 110,000 acres of the original ranch could not attract the numbers of visitors hoped for by FWS planners. Acquisitions along Arivaca Creek and in Brown Canyon have significantly expanded public use opportunities, but at the expense of distracting refuge operations from the challenges of masked bobwhite reintroduction. The idea that the Buenos Aires Refuge is a "natural" place, presumptively beneficial to all wildlife, has served to obscure the trade-offs necessitated by limited budgetary resources and unanticipated ecological phenomena.

The Buenos Aires Refuge, in common with a great deal of environmental activism, disproportionately reflects the aesthetic and recreational expectations of wealthier Americans. This fact has been masked, however, by the superficially benevolent, unobjectionable ambition of wildlife preservation. It is as though the refuge were attempting to accomplish in the ecological realm what the welfare state previously envisioned for the economy: a safety net for those who fail to compete successfully, intended to restore them to the wilds of the market so that they can survive unaided.

Evaluated scientifically, the refuge remains largely a symbolic achievement. There is little evidence that FWS ownership and management have attracted any new wildlife to the Buenos Aires Refuge. If anything, the drying up of Aguirre Lake and other water sources has probably diminished the biological diversity of the ranch since 1980. It is by making wildlife species and the land they inhabit into *public* property that the FWS reproduces the bureaucratic capital it has invested in the Buenos Aires Refuge. Acquisition of riparian areas, increased documentation of wildlife sightings, and the attraction of *humans* to the refuge have together created the (mis)perception that the FWS is responsible for the presence of wildlife species that, with few exceptions, were already there.

counterfeiting conservation

Dramatic environmental change in the Altar Valley has resulted from complex interactions among numerous natural and social processes: grazing, rainfall and drought, range management and improvements, erosion, market conditions, fire, and fire suppression. Generally the greatest changes occurred when multiple factors coincided—as with grazing during the droughts of the 1890s, 1920s, and 1950s—or when a particular sequence of events pushed natural systems across critical thresholds: water development, followed by overgrazing during drought, followed by torrential rains at the turn of the century, for example. These interactions were nonlinear in their effects, and simple causal relations are probably impossible to establish, even negatively. Would the Altar Wash have formed in the absence of Aguirre Lake? Quite possibly yes. In the absence of grazing? Maybe, maybe not. Had there been no cattle, would the masked bobwhite have survived the droughts of the 1890s? No one knows. Even mesquite encroachment might have occurred in the absence of grazing and other human impacts because of long-term shifts in the temporal distribution of rainfall (Swetnam and Betancourt 1998).

Whatever their causes, two changes stand out for their impacts on the structure and functioning of the Altar Valley ecosystem. First, arroyo formation fundamentally rearranged the valley's hydrology, replacing broad, relatively moist bottomlands with drier terraces and sandy channels. Over the years, headcutting from the Altar Wash has progressed for miles up tributary drainages, except on one ranch, where protective dikes have been assiduously maintained by three generations of a single family. The alluvial bottomlands—once the most productive and lush areas in the valley and the core habitat of the masked bobwhite—are shadows of their former selves. Yet very little attention has been paid to the efforts of ranchers, such as the Gills and the Victorio Company, to mitigate entrenchment through spreader dams, dikes, and grass seeding. These measures, although not uniformly successful, indicate both the potential for healing arroyos and the delicate balance between upland vegetation, grazing, and runoff.

Second, conversion of grassland vegetation communities to shrub-dominated ones has had negative impacts on hydrological functioning, erosion and infiltration, certain wildlife, and forage production. In the Altar Valley, the shift occurred later than elsewhere in the region; up to the late 1930s and early 1940s, perennial grasses remained dominant in the wetter, higher end of the valley where the Buenos Aires is located. The lag may be attributed to a combination of factors: later fence construction, allowing fire to continue into the 1920s; the summer rest from grazing reported in the 1920s by C. B. Brown; and the depressed mesquite populations following fuelwood harvesting in the 1890s. Although shrub establishment must have begun earlier, the drought of the 1950s appears to have been the threshold event, beyond which a natural return to grassland was impossible.

For the better part of the twentieth century, ecological theory held that rangeland vegetation change would reverse itself naturally if livestock grazing were curtailed or eliminated. For early ranchers and agency researchers, this seems to have been intuitively obvious; the work of Clements, Sampson, and Dyksterhuis formalized it and helped to institutionalize it in range management. It remains a widely held belief among the general public today. Coupled with fire, it is the fundamental premise of management on the Buenos Aires Refuge. The Clementsian model, however, is no longer accepted among ecologists studying semiarid plant communities (McPherson and Weltzin 2000; Westoby et al. 1989), for reasons that the Buenos Aires case illustrates. Stocking rates declined for most of the century, yet vegetation change continued almost unabated. Massive, expensive manipulation of vegetation was necessary to restore grassland of any kind, whereas livestock exclusion, even in combination with fire, has had no demonstrable effect on the composition or density of vegetation since the creation of the refuge. Arroyo formation and the shift to mesquite dominance were threshold events, disrupting any neat, linear relationship among seral stages. "Thus, although livestock grazing (particularly in combination with other factors) played an important role in vegetation change shortly after Anglo settlement, excluding livestock from most sites now will have little or no impact on abundance of woody plants or non-native herbs during the next several decades" (McPherson and Weltzin 2000: 4).

Ecological restoration in the Altar Valley must address arroyos and mesquites. Neither can be "healed" simply by curtailing the activities that contributed to degradation in the past because both the activities and the environment have changed in fundamental ways. The successes that have been achieved have proved temporary and, since about 1970, economically untenable. Spreader dams healed the arroyo on Arivaca Wash, but the Altar Wash is too large for ranchers to afford to tackle, given the economics of cattle production. Bulldozers and aerial seeding restored grasses on much of the Buenos Aires in the 1970s, but these tools proved neither lasting nor economical over time. Fire is clearly necessary, and it may suffice if applied regularly over a period of many decades. Where fine fuel loads are inadequate, however, fire must be preceded by more intensive treatments. The introduction of Lehmann lovegrass further complicates this approach. No other species has proved nearly as effective in restoring perennial grass cover in the area, and hence, fuel for fires as well as forage for livestock. Studies to date, however, have consistently found that Lehmann lovegrass withstands fire better than native perennial grasses do (Cable 1965, 1971; Robinett 1994; Laura Huenneke, personal communication, 12 July 2001), meaning that burning areas where Lehmann lovegrass is established may indirectly harm the native species that it is intended to benefit. Moreover, topsoil losses in upland areas may have been so great that native perennial grasses cannot re-establish their earlier densities on any human time scale. Above all, it must be stressed that science does not yet understand these ecosystems with precision. We simply do not know how to control them; if we did, they probably would not be there anymore.

CONTESTING THE REFUGE

Public debates over the Buenos Aires Refuge have rarely raised these historical and ecological issues. Instead, partisans have focused on cattle and the masked bobwhite, each side upholding one and disparaging the other. It is surprisingly difficult to reconstruct the debates in historical detail. Newspaper articles are invariably brief and sometimes distorted;

with a handful of exceptions, FWS documents are no better. The Annual Narrative Reports, for example, strongly resemble Spanish missionaries' *cartas annuas* to their far-off superiors: short on self-examination, long on self-praise, contemptuous of the natives, and replete with pleas for more funds. The oral accounts I have collected are riddled with inconsistencies. In some cases, people have taken away opposing impressions of the same event. A great deal of personal animosity has developed, which need not be aired here. Instead, I seek to situate the disputes over the refuge in the larger context described in the preceding chapters.

The most consistent feature of debates about the refuge has been a demographic one: Critics have almost invariably been rural residents, whereas supporters' ranks have been dominated by people from Tucson, Phoenix, and other urban communities. Refuge management has exacerbated this division by reaching beyond the local community for visitors and political support. The most outspoken critics have come from multigeneration ranching families organized in the SACPA, a membership organization founded in the 1920s to "protect" against rustlers. As mentioned in chapter 8, the dispute flared up in the mid-1990s on the heels of heightened media coverage of refuge public use developments.

In July 1996 SACPA and the Society for Environmental Truth[1] copublished a 78-page report on the refuge (SACPA 1996), distributing it to FWS officials and elected representatives. Sympathetic legislators at the state level were moved by the report to introduce a "memorial" (a sort of sense-of-the-body motion to be delivered to the U.S. Congress) calling for an Office of Management and Budget investigation of the refuge. The motion passed the House but failed in the Senate after a flurry of phone lobbying by the Arizona League of Conservation Voters. In the end, the report did little more than cement the opposing arguments of refuge supporters and detractors. Its importance lies not in its effects but in the light it sheds on the two sides' ways of evaluating the refuge.

The core argument of the SACPA report was that the masked bobwhite program is a failure, probably for reasons beyond the control of the FWS (e.g., inadequate summer rains or excessively cold winters). A fairly thorough review of the scientific literature and refuge documents was presented in support of this contention (SACPA 1996: 19–28). The report implied that the FWS has redefined refuge objectives toward more general goals of habitat preservation, as evidenced by its aggressive land acquisition program, in order to cover up these failures. The refuge suffers from "mission creep" (ibid.: 60).

SACPA's criticisms of the masked bobwhite program were largely valid, as we have seen, although there is no way to know definitively that the program is a failure. The FWS's response on this point was simply to assert the opposite, placing heavy emphasis on the exclusion of livestock for the alleged success of reintroductions. Against SACPA's claim that masked bobwhite population surveys are inaccurate and perhaps exaggerated (SACPA 1996: 25), the refuge presented anecdotal evidence of limited reproduction in the wild (FWS 1996: 5); both sides may be correct because their claims are not mutually exclusive. The debate, however, was not really about the science of masked bobwhite reintroductions.

The SACPA report constructed the masked bobwhite on the model of commodity production, indicting the refuge as a waste of taxpayers' money. An entire chapter was devoted to calculating "the taxpayers' investment in the BANWR," with lost opportunity costs (what the ranch would have contributed to the regional economy, complete with a "multiplier effect") and interest (because the federal government was running a deficit at the time) added to actual expenditures. All told, the report concluded, the Buenos Aires Refuge has cost taxpayers more than $68 million.

> To get some perspective on this cost, if a population of 100 MBQ [masked bobwhite quail] on the refuge is assumed, the total amount invested per surviving bird at the end of 1995 is more than $687,000. If the population is assumed to be 300 birds then the investment is $229,000 per bird. If only the operation and maintenance costs are considered the cost is $93,250 per bird. If a population of 300 is assumed, than [sic] the cost per bird drops to $31,080 per bird. At an average weight of seven ounces for an adult quail, that is $4,440 per ounce or more than 11 times the value of gold. (SACPA 1996: 47)

It hardly needs to be pointed out that this way of looking at the masked bobwhite exactly reproduced the manner in which ranchers' calves and steers are evaluated in the market: weight times price equals value. The masked bobwhite thus appeared as a commodity, with the taxpayer as consumer: Would you pay this much for this product? The metaphor of "solid gold quail" dominated local reception of SACPA's report, overshadowing the more scientific questions surrounding reintroduction.

Such an approach could only strike refuge supporters as crass, if not outright offensive. Putting a price on an endangered species was to them sacrilegious. The refuge had no solid evidence to refute SACPA's basic claims about the masked bobwhite program, but the larger rhetorical thrust of the report made refutation unnecessary. Instead refuge officials and supporters focused on the cattlemen's imputation of a federal conspiracy "to impose . . . government control of the land and water" (SACPA 1996: 60) and interpreted the entire report as nothing more than a poorly disguised attempt to open the refuge to grazing. The report made no such demand, but prior events confirmed them in their suspicions. The Society for Environmental Truth had been publicly discredited as a shill for industry in an *Arizona Daily Star* editorial the year before (18 April 1995), and its executive director had written numerous articles containing exaggerated or misleading claims about the refuge. In late 1995 SACPA's president had proposed a grazing program for the refuge, falsely claiming that cattle would eat noxious shrubs to the benefit of perennial grasses.

When the NBC *Nightly News* featured the Buenos Aires Refuge in its "Fleecing of America" series on government waste in January 1996, giving SACPA's claims broad exposure, supporters reacted with a wave of angry letters to local newspapers, NBC, the refuge, and elected officials. Judging from the letters, one can surmise that refuge supporters hold the opposition between cattle grazing and wildlife habitat—more social than ecological—as an article of faith. They are apparently persuaded that by removing livestock the FWS has restored the Buenos Aires to its "original" Sonoran savanna grassland conditions. As a contribution to the latest management plan effort, a long-time member of the Washington Fish and Wildlife Commission wrote, "The Buenos Aires Wildlife Refuge is an outstanding example of restoration of land which had been overgrazed for many years. The proper management of the last 12 years has revitalized the Sonora Desert to what it was 100 years ago. Further, the absence of cattle on the land has resulted in an area teeming with birds, deer, quail and other wildlife" (FWS 1998). These claims are quite simply false, but they recur in refuge reports, newspaper and magazine articles, and conversations among visitors to the refuge.

There are two ways to explain how this misapprehension of the Buenos Aires Refuge has become so prevalent. One is ignorance: Most people simply do not know better. Some people, however, especially refuge employees, do know better. The other explanation is wishful thinking: People want to believe that an "original" piece of Arizona still exists, that some place has been spared or restored from the damages of overgrazing. Charles Bowden (1992: 40, 43, 47) constructs the Buenos Aires Refuge as "a place outside of time. . . . All around you spreads the landscape that greeted the first padres and conquistadors. . . . The sweep of grass is something that was here before recorded time and will be here after we're gone." He suggests that people visit the refuge "to escape a sense of time. . . . In the valley, all the obvious signs of civilization—the power lines, neon signs, railroad tracks, gridlocked cities—all this finally ends." Another magazine echoes this view with the cover headline "Seeking Refuge: Endangered Wildlife and Weary City Dwellers Find Safe Haven at Buenos Aires Preserve" (Stocking 1995).

The same oppositions that gave rise to the mythology of ranching a century ago now make possible the mythology of the refuge. Ordinary life is dominated by work, coordinated by the clock, and characterized by artifice and congestion; the refuge, like ranching for Teddy Roosevelt, is the opposite. Only now there must be no sign of any commodity production whatsoever; all traces of human labor must be swept or wished away. Where industrial capitalism romanticized pastoralism as a kind of noble, primitive work, the "New West" economy valorizes recreation, leisure, and places dedicated to "nonconsumptive" consumption. It is as though, having decoupled the economy from natural processes of rainfall and vegetation growth, suburban life conjures a desire for unpeopled, "pristine" Nature outside of the cities. It is here that the conflict between ranchers and the refuge is most absolute. For ranchers, the distinction between work and leisure does not apply. Their place of work is also their home, where they relax and socialize; their days are not divided into fixed hours of labor and idle hours of recreation. For them, nature is a site and source of production, not its compensatory opposite. Refuge employees receive their salaries regardless of whether the masked bobwhite or other wildlife thrive because it is understood that natural processes cannot be wholly controlled or guaranteed. Ranchers struggle with the same unpredictability but with direct financial consequences. In opposition to the institutional security of the refuge, they valorize the risk involved in "putting our necks out" and "signing that note across that desk in the bank." The refuge manager, one rancher exclaimed, has no idea of "the sweat and tears that have gone into building this ranch up. He hasn't got anything in [the refuge] but his lunch bucket!"

The only groups of people who might be able to break this deadlock are biologists and ecologists, who have knowledge and authority that both sides claim to respect. But the academic division of wildlife from livestock, ecology from range management, and nature from the economy recapitulates the social division between environmentalists and ranchers. Interviews conducted in 1998 with former masked bobwhite researchers found widespread criticism of the refuge's biology program. "The program in my opinion is not good at all," said David Brown. "The survey methods are illogical, the research is poor, and the continuation of releasing pen-reared birds is keeping it from success. . . . The articles they've submitted are appalling." There is no consensus as to what should be done, however. Brown contends that birds should be captured in Mexico for release;[2] Steve Dobrott and the Levy brothers suspect that winter feed is lacking, particularly white-ball acacia (Acacia angustissima); David Ellis believes that there are too many mesquite trees, which favor Gambel's quail, and that predation by raptors is a problem. In general, scientists have refrained from public criticism out of concern that their words might play into the hands of the cattlemen or damage their professional relations with the refuge. Science on the refuge remains subordinate to politics.

At the time it was made, SACPA's second charge against the refuge—of mission creep—was difficult to prove or disprove. More recently, though, the FWS has effectively inculpated itself. The Draft Comprehensive Conservation Plan for the Buenos Aires Refuge (FWS 2000), released in December 2000 after nearly three years' preparation, asserts repeatedly that "the primary reason the Refuge was established" was not masked bobwhite recovery but restoration of the Sonoran savanna grassland habitat type. This falsehood, which is contradicted by FWS documents dating back to the first Masked Bobwhite Recovery Plan, is then used to justify substantial staff expansion and a wide variety of new programs and capital improvements. Because of the function of paper representations in bureaucratic reproduction, this piece of revisionist backpedaling takes on real significance. If it survives into the final plan, it will give the FWS an official basis for abandoning masked bobwhite restoration without calling into question the continued existence of the refuge itself.

THE COSTS OF SYMBOLIC ENVIRONMENTALISM

Judged against the historical record, the debate over the Buenos Aires Refuge has been quite simply beside the point. Neither cattle nor the masked bobwhite have that much to do with ecological restoration in the valley. Of the fifty-nine masked bobwhite reintroductions attempted prior to creation of the refuge, the only demonstrated successes occurred in the presence of cattle grazing. Several conducted in ungrazed areas failed. Releases in the bottomlands of the Buenos Aires Ranch prior to the Victorio Company's bulldozing were unsuccessful (Brown and Ellis 1977); afterward they succeeded, at least until drought struck. Ever since floods wiped out the ranch's spreader dams in 1983, released birds have again struggled to reproduce, except in years of very good summer rainfall. If one considers the role of microclimatic humidity levels in triggering breeding, this correlation should be no surprise. Simple induction suggests that the obstacle to masked bobwhite restoration is not cattle but habitat change and that the management practices implemented by the Victorio Company to improve grazing conditions also benefited the masked bobwhites released in the late 1970s. Ironically such practices are more difficult to implement now than before because of limited refuge resources and greater regulatory restrictions on federally owned land.

Conversely there is little biological reason why ranching and endangered species cannot coexist. The same basic ecological processes support both wildlife and livestock, after all (Sayre 2001). Millions of dollars of public research, government subsidies, and ranchers' investments have aspired to reduce these processes to industrial-style control, but it cannot be concluded that they have succeeded. The pastoralist basis of range livestock production—reliance on nature to do the value-producing work—has stubbornly refused to submit to technological quick fixes. That the western range retains as much biodiversity as it does—compared, for instance, to most farms in the United States—is a consequence of this very fact. It is time to acknowledge that dominating the range has failed (Holling and Meffe 1996; Nelson 1995) and that restoration will be realized only by embracing, rather than trying to eliminate, natural variability and the biological diversity that has evolved to cope with it.

The principal threat to ranching is economic: declining returns on the one hand and rising opportunity costs on the other. Corporate concentration in the beef industry is even greater now than a century ago: Like the narrow center of an hourglass, three huge companies dominate market relations among thousands of producers and millions of consumers (Heffernan and Constance 1994; Skaggs 1986). These firms seem determined to reorganize the production of cattle on the same model as poultry (and, for that matter, captive masked bobwhites): large numbers of animals confined in small areas, with their growth, health, and reproduction maximized, minutely controlled, and as divorced from natural processes as possible. In recent years the model has been applied to pigs, with devastating effects on rural producers, landscapes, and communities (Thu and Durrenberger 1998). It has not been achieved with cattle yet, but technological innovations in breeding, artificial insemination, and embryo transfer, as well as the routine use of antibiotics, vaccines, and hormones, have already helped to undermine ranching by creating advantages for non-range-based producers.

The pressures of suburbanization are also eroding the foundation of range conservation: ranchers' economic incentive to manage for the long term (Workman and Fowler 1986). The flood of outside capital into western real estate recapitulates the cattle boom, replacing livestock with people, livestock loans with home mortgages, and stock tanks with subdivisions. With or without prior intent, ranch owners are thrust into the role of real estate speculators, compelled to defend their equity in terms of development potential even if they hope never to take advantage of it. Only the high symbolic capital of being a rancher has kept nonspeculators in the business. Once promoted by the state as a means of integrating marginal lands into capitalist markets, ranching now finds itself the object of increasingly rigorous and complex state regulation, economically marginalized and ecologically criminalized. Returns are no longer adequate to cover the high costs of intensive range restoration, and the important work of managing land well suffers from the growing costs in time and money of fending off lawsuits, attending public meetings, and coping with recreationalists, undocumented immigrants, and the Border Patrol.

NATURALIZING SUBURBIA

On 13 January 2000, Tucsonans awoke to a front-page feature story about the Buenos Aires Refuge in the *Arizona Daily Star*. Under the headline "Back Home, Back Home on the Range: 88 Antelope Play Once Again at Buenos Aires Refuge," the article presented a heartwarming story. "When this patch of Sonoran savanna was designated as a haven for an endangered species 15 years ago," it began, "a century of cattle grazing and hunting had left it parched and eerily quiet." Readers were led to understand that livestock exclusion and fire were reversing this and that the transplanted pronghorn were part of the refuge's larger ambition "to restore the grassland ecosystem that originally was found here," in the words of the refuge manager. Once, he said, the valley had been "bounding with antelope as far as the eye could see."[3]

Embedded in the story, the principal reason for the relocations received less emphasis. The pronghorns had been captured in the Prescott Valley northwest of Phoenix from the state's largest and healthiest herd in order to get them out of the way of imminent suburban development. A ranch was being subdivided. "During the roundup, wildlife officials found water mains and manholes in grassy hillsides, signaling that new buildings and roads will soon follow." An AGFD official described the transplant as "a good opportunity for us to pull some animals out of a population that's doing exceptionally well before development starts to ruin habitat."

The story of the pronghorn relocation encapsulates the role that the Buenos Aires Refuge plays at a regional level, where mediated representations outstrip facts on the ground. What would otherwise be a tragic, even scandalous story of suburban sprawl ruining prime habitat for a large, charismatic mammal becomes instead a story of redemption: an ecosystem putatively being put back together. The released pronghorns quickly dispersed in all directions—few of them remained on the Buenos

Aires Refuge as little as a week later—but no follow-up story appeared in the paper. The misleading characterization of the refuge's history went unquestioned and uncorroborated. The facts, it seems, were unimportant compared to the implicit message: Subdivisions are not really a problem because the wildlife have a refuge on the Buenos Aires.

This is a message that few environmentalists or wildlife ecologists would openly embrace, although some do (Donahue 1999; Wuerthner 1994). It is not so much a position as a rationalization, a way of casting the status quo in a positive light. Those who seek to end grazing on public lands, however, cannot disclaim all responsibility for the collateral impacts of their activism: for example, the accelerated subdivision of private lands presently in ranching (Rowe et al. 2001). In the politics of land use and conservation, moreover, "preserved" areas such as the Buenos Aires Refuge serve to "mitigate" (in the jargon of the FWS) the ongoing fragmentation and development of the rest of the landscape. Ranch improvements on the Buenos Aires were devalued economically when grazing ceased; some were physically removed, and others simply deteriorated with time. Nevertheless, the FWS capitalized on the Victorio Company's investments by shifting the mediation of value on the Buenos Aires Refuge. The beneficial aspects of the ranch period—from Aguirre Lake to revegetation to erosion control—were appropriated symbolically as "natural" and thereby ideologically erased; anything deemed negative—mesquites, Lehmann lovegrass, and arroyos—was attributed to cattle ranching. The wildlife itself underwent a similar transformation from chance occupant to symbolic commodity. Socioculturally these are precisely the steps by which other ranches in the region have been incorporated into a new economic order, with one exception: The Buenos Aires services recreational rather than residential purposes.

Constructing the Buenos Aires Refuge as "natural" is critical to this ideological project. The refuge issued from the symbolic opposition of the masked bobwhite to cattle ranching, the dominant land use of Arizona's past. Its reproduction as a refuge has depended, thus far, on the widespread misperception that removing cattle "heals" the damage of overgrazing "naturally." The appeal of this idea is twofold: First, it resonates with venerable ecological theories that are taken for granted by many urban Arizonans; second, it implicitly legitimates the dominant land use of the present—suburban development—on which those same people depend economically. Urbanization can appear environmentally benign if accompanied by the elimination of cattle grazing on outlying lands now valued more highly for recreation than for their capacity to produce food. The impulse to produce a habitat for leisure, once confined to the suburban lawn, has expanded to encompass all public lands. By appearing to invert both ranching and the urban landscape, the Buenos Aires Refuge resolves on a symbolic level conflicts that have yet to be resolved in more substantive ways.

THE FUTURE OF THE ALTAR VALLEY

The masked bobwhite is no longer Arizona's most famous bird. That title has passed to other endangered species. In some areas it is the southwestern willow flycatcher, in others the Mexican spotted owl. In southern Arizona, it is unquestionably the cactus ferruginous pygmy-owl *(Glaucidium brasilianum cactorum),* a subspecies that is abundant in Mexico but quite rare in Arizona, which is at the northern edge of its historical range. Pressured by environmental groups, the FWS listed the Arizona population as endangered in 1994; litigation forced designation of critical habitat in late 1998. A large chunk of the critical habitat lies in northwest Tucson, ground zero for suburban sprawl in the 1990s; much of the rest is in the Altar Valley.

Fearful that its permitting actions for roads, subdivisions, and other projects would expose it to liability under the Endangered Species Act, Pima County launched the Sonoran Desert Conservation Plan almost immediately after critical habitat for the pygmy-owl was designated. The plan seeks to obtain a permit from the FWS that would allow a certain amount of impact to endangered species in return for comprehensive, long-term habitat conservation. For nearly four years now, county officials, scientists, consultants, and members of the public have been working to develop the plan.

As surveys for pygmy-owls have been expanded and scientific information on other species has been assembled, the importance of the Altar Valley for regional conservation has grown dramatically. Its relatively unfragmented landscape harbors about as many pygmy-owls as northwest Tucson; most of the Altar Valley owls inhabit ranches, not the Buenos Aires Refuge. Of the "priority vulnerable species" of concern to planners, more are found in the Altar Valley than in any other watershed in eastern Pima County. Its geographic location—linking Mexico with the Tucson area and with lower desert to the north—heightens its significance for conservation of pygmy-owls and a variety of migratory species. County officials and planners have begun to see the Altar Valley as critical to the plan as a whole: If it can be spared from subdivision, it may serve to mitigate development impacts all around Tucson.

The question is how to achieve this. Ranchers in the valley favor conservation of present land-use patterns: that is, continued ranching. They know that subdivision would greatly restrict the possibility of fires, which they wish to use to help restore grasses, and that housing development on the old floodplain, where much of the private land is located, would preclude restoration of the Altar Wash. They also know, however, that their property values reflect the potential for suburban development, and they reject any measure that would eviscerate their equity though zoning or unilateral land-use restrictions. The county has pledged to make "ranch conservation" a central component of the plan, and there is reason to hope that a program to purchase development rights from ranch owners will emerge. This would remove the pressures of real estate appreciation and restore the economic basis for long-term stewardship.

The Sonoran Desert Conservation Plan is fraught with complexities and potential pitfalls. Here is a major one: What is the real criterion of success? Is it actual conservation of species on the ground or the prevention of further lawsuits? In theory, these ought to be one and the same. As the Buenos Aires case demonstrates, however, no such identity can be assumed. The largest population of pygmy-owls in the Altar Valley, for example, occupies a pasture that was burned in the 1980s as a range improvement measure; yet the FWS sees fire as a threat to the subspecies. The area has been grazed for more than a century, and large numbers of pygmy-owls have been found inhabiting severely overgrazed areas in Mexico; yet grazing, too, is viewed with suspicion. It is as though the issue were not the actual wildlife but simply legal liability: If the FWS does not permit anything, then it cannot be sued. This pattern of science subordinated to politics has also been documented elsewhere in the region (Sheridan 2001).

The lessons of the Buenos Aires Refuge must be recognized and applied if conservation is to be achieved in the Altar Valley. Subdivision knocks at the north end of the valley; its entrance would represent a new threshold crossed, foreclosing options for restoration while triggering the cycle of land appreciation and fragmentation. Even the refuge's staunchest critics recognize one virtue: It effectively prevents suburban development on a large core area of the valley. Expanding this to the watershed scale will require a coalition of scientists, agencies, environmentalists, and above all, local ranchers. Building such a coalition should begin with a candid assessment of the refuge.

For decades environmentalists have decried the co-optation of public agencies by commercial interests, the distortion of science to justify economic exploitation. They have correctly pointed out the damage done by cattle grazing across much of the West. Science, however, can be distorted to serve any interest group, including environmentalists, and the remedy they propose is at best naïve. How the damage occurred and what will heal it are two entirely separate questions. Those who assert that livestock exclusion will "save" endangered wildlife species such as the masked bobwhite or the pygmy-owl raise false hopes among the general public while unnecessarily antagonizing local residents. Today the ascendant land use seeks to exploit public lands for recreation and as scenic backdrops and outsized backyards for suburban residents. Its impacts are more difficult to manage or measure and its beneficiaries are more difficult to demonize than those of the past. The fetishized category of endangered species—created by the state, modeled on the commodity, and infused with racial essentialism—is a narrow and distorting lens through which to survey the challenges ahead.

notes

INTRODUCTION

1. In some places, Jordan insists that ranching is an Old World phenomenon. "Ranching was not a product of the American frontier . . . or of the semiarid West. . . . Nor did it arise in Latin America" (1993: 14). In other places, however, he seems to avoid calling Old World livestock husbandry "ranching," using instead the phrase "range cattle raising" or "the open-range herding of beef cattle" (ibid.).

2. Three platform mounds have been found on the pediment at the base of the Coyote Mountains in the northwest corner of the Altar Valley. They have not been excavated, but presumably they date to the Hohokam period or earlier (Dart et al. 1990).

1 THE SOCIAL PRODUCTION OF THE ENDANGERED MASKED BOBWHITE

1. According to Whitaker (1986: 35–36), zoologist Griffing Bancroft and rancher Les Woodell, on a trip to look for masked bobwhites in the Magdalena area in 1927, found them in only one place: in cages behind a restaurant, where they were being served as an entrée. I have no reason to doubt the truth of this account, but it does not appear in any reviewed sources and may be apocryphal.

2. This construction is unquestionably a product of retrospect. In the 1870s, Brown could not know that the bird would be extirpated from Arizona in twenty-five years; by 1904, this fact definitely colored his presentation.

3. I am indebted to Professor Steve Russell for helping me to understand the history and principles behind masked bobwhite classification. See Brown (1989: 123–142) for a more detailed discussion.

4. Tomlinson (1972b) gives the fullest account of this point. The sightings east of the Santa Cruz Valley were ruled out as fraudulent or as mistaken Mearns' quail sightings. Most scientists now believe that the higher elevations of the valleys in question rule out previous occupation by masked bobwhites.

5. Ranchers obstructed early game laws, according to Brown, thereby abetting the game hunters. "In 1887, I think, the first game law was introduced in the territorial legislature. The bill originated in the Tucson Gun Club, and its purpose was largely the protection of 'Quail,' but so great a pest were the birds regarded by the ranchmen in the Salt River valley that the legislators from Maricopa County threatened to kill the bill unless the clause protecting 'Quail' was stricken out" (Brown 1900: 33).

6. Jim Levy, personal communication, 20 March 1998.

7. "The standard quail drive-net consists of a catch tube or barrel 25 feet long, made of 1-inch mesh twine netting supported by round and semi-round hardwood hoops, and of two wings each 150 feet long and 12 inches high. . . . To complete and spread the net to the desired dimensions, 18 semi-square and round hoops are placed within, approximately 15 to 18 inches apart. The first of these, forming the entrance, is 15 inches wide by 14 inches high; the others by slight graduations reduce to 12 inches wide by 10 inches high for the fifth hoop. . . . The two wings of the net are completed by placing $1/2$- or $5/8$-inch round hardwood stakes in the web at intervals of 5 feet near the net entrance, and 6 feet farther out. These stakes are 15 to 16 inches long, and pointed for driving into the ground" (Ligon 1946: 33ff.). These selections are provided to illustrate the tone and detail of Ligon's instructions, which are very much longer than space allows here.

8. Jim Levy, personal communication, 20 March 1998.

9. In 1936 or 1937, AGFD Deputy Game Warden Warren Peterson reported sighting a covey of masked bobwhites in Jalisco Canyon east of Arivaca, and AGFD officials appear to have treated it as credible (Arrington 1942). David Gorsuch also reported seeing masked bobwhites in the area in 1937 (Tomlinson 1972b: 13). Tomlinson (ibid.) dismissed the sightings in 1972, however, writing that although both men "definitely knew masked bobwhites . . . , these observations were never substantiated. Even if correct, the sightings probably represented a remnant population that has since died out. For all practical purposes, then, masked bobwhites had disappeared from Arizona by the early 1900's." Tomlinson's distinction between scientific and practical purposes is far from clear. By 1972 it may have appeared advantageous to discredit the 1936–37 sightings because they did not accord with the view that cattle grazing had dealt a swift and decisive blow to masked bobwhites in the United States and would soon do the same in Mexico.

10. David Brown, personal communication, 20 May 1998.

11. Tomlinson (1972b: iv), after conducting field and literature surveys for the FWS for four years, concluded that the masked bobwhite "probably never was a widespread and abundant bird."

12. "By sheer chance we located a population of this race [masked bobwhite] 200 miles north of Guaymas in an area where none had previously been reported and where we had no reason to expect any" (Gallizioli et al. 1967: 572–573). It is not true, however, that no one had ever found masked bobwhites on Rancho El Carrizo before; that area was the source of the last specimens collected for museums in 1931 by J. T. Wright (Tomlinson 1972b: 9). It is not known if Ligon knew

of Wright's find; the Levys learned of it only later (Seymour Levy, letter to author, 17 April 1999). In any case, the element of miraculousness in the "rediscovery" must be viewed as part of the larger mythologization of the masked bobwhite, akin to the present refuge manager's habit of telling visitors and journalists that the entire captive population is descended from eight birds, a claim that is tantalizingly dramatic but factually false.

13. This legal division of wildlife-related labor was established in 1913 under the Weeks-McLean Act on the reasoning that migratory birds are a type of interstate commerce (Dunlap 1988: 37).

14. One of Ligon's partners in the 1949 and 1950 trips to Sonora, for example, was L. L. Lawson of the Arizona Federal Aid Service. Lawson authored many reports about game birds for the Arizona Game and Fish Commission in the 1940s and 1950s (see Brown 1989: 288–289). In 1952 Ligon wrote that a "cooperative plan" between the Arizona and New Mexico Game and Fish Departments "has at last resulted in prospects for saving the birds from extermination," although this plan does not reappear in the literature. In 1964, as noted already, the Levy brothers were accompanied by an AGFD biologist. In the same year, the U.S. Bureau of Land Management leased some land to the Arizona-Sonora Desert Museum for an unsuccessful attempt to reintroduce masked bobwhite in the Altar Valley (Gallizioli et al. 1967: 572).

2 THE CATTLE BOOM IN THE ALTAR VALLEY

1. Aguirre may also have been prompted to move by the development of a well at Sasabe between 1857 and 1862, which allowed the road to Altar, Sonora, to bypass Presumido Canyon entirely.

2. Records of the Arizona Department of Water Resources. The State of Arizona did not require registration of surface water rights until 1919, at which time Jack Kinney filed papers on all the tanks on La Osa's huge holdings, backdating all of them to 31 December 1883. Consequently it is impossible to determine with any more precision when the tanks were built.

3. General Land Office (GLO), claim nos. 266 (27 April 1891), 1008 and 1009 (27 March 1893), and 710 and 711 (30 November 1904).

4. The name appears sometimes as Russel and sometimes preceded by "Dr." or "Derrick."

5. An article from the *San Francisco Bulletin,* reprinted in the *Arizona Daily Star* (10 February 1887, p. 4), credits Hemme with having constructed the waterline and a deep well, "preparing to grow crops of various kinds, particularly feed for cattle. He has brought horses and many high grade, short horn cattle from his home farm in Contra Costa county [California]." These claims are not implausible,

but there is reason to believe they are exaggerated. Improved breeds were much commented upon at the time, but this herd does not appear anywhere else in the literature. Nor is there any other evidence that the waterline was constructed, though a similar one was installed sometime between 1915 and the early 1920s by the La Osa Live Stock and Loan Company (see chapter 3).

6. The density may be estimated from a number of instances in his notes, in T 18&19S, R9E, where, standing at a single point, he measured the bearing and distance to mesquite trees. The farthest he measured was about 250 links, suggesting an areal measure of about two acres; he listed one to four mesquites at each point. That he gave diameter measurements suggests that the trees were single-trunked and stood upright, unlike the bushy, multiple-trunked trees typical of more recent years.

7. The line ran north-south between ranges 10 and 11 east and from township 6 to 15 south, or roughly from present-day Marana to Three Points.

8. *Noon* v. *Bogan,* 7 December 1908. I would like to thank Mary Noon Kasulaitis for sharing this transcript.

9. Unfortunately, the Altar Valley received only occasional attention from the SCS in its studies of the Santa Cruz watershed. Because no extensive farm fields were in the valley, downcutting there did not seem to qualify as "damage" (CSWR MSS 289, Box 6, Folder 14, p. 15).

10. Two items in Cooke and Reeves (1976: 58) are in error. First, Bryan (1925: 109) does not specify the length of the entrenched channel; the claim that it extended "almost to the Anvil ranch" is a mistaken reading of Bryan's description of the bluffs of upper alluvium adjacent to the bottomlands area. Second, Cooke and Reeves cite Andrews as reporting an arroyo twenty feet deep and up to six hundred feet wide; these figures, although plausible, appear nowhere in Andrews's text.

11. "For animals to become capital for their owners, they must be bought as well as sold" (Ingold 1980: 231).

12. Leases were earlier instituted for grazing on state lands in Texas under laws passed in 1879 and 1883, with important precedents for the larger western range. See Coville (1905).

3 THE FORMATION OF RANCHING

1. In an interview with Yginio Aguirre in 1986, FWS archaeologist David Siegel seems to have considered that the refuge should revert to the "original" spelling, in keeping with its objective of restoring the land to its "original" condition.

2. Unless otherwise noted, all information on Jack Kinney and the La Osa

period is derived from Kinney's papers, housed at the Arizona Historical Society in Tucson.

3. The ostensive reasons for this state-capital alliance will emerge in the pages that follow. It must be kept in mind, however, that range reform was but one aspect of a larger pattern in the American West that encompassed federal agency regulation and development of water resources (Reisner 1986; Worster 1985), timber (Hirt 1994), and public lands in general (Hays 1959; Stegner 1954). Efforts to improve range conditions, for instance, were partly motivated by the need to control erosion, which threatened agricultural production, municipal water supplies, and newly erected dams in the early twentieth century.

4. By the original Enabling Act (61st Cong., 2d sess. [20 June 1910], ch. 310, sec. 28), five years was the maximum term that could be leased without competitive bidding. This was later amended to ten years (74th Cong., 2d sess. [5 June 1936], 517).

5. Karl Ronstadt, personal communication, 18 August 1998.

6. The parallel with social evolutionist theories—which claim that societies "naturally" evolve from savagery through pastoralism and agriculture to industrial civilization—should be obvious.

7. It is no coincidence that money, viewed from the vantage of finance capital, shares most of the qualities Ingold derives as essential to "the unit of wealth in a pastoral economy": convertibility into consumable products, (the appearance of) "natural" self-reproduction, private ownership, and ease of aggregation (1980: 227).

4 PRODUCING NATURE

1. According to their hired Buenos Aires foreman at that time, whose draft deferment depended on this fact.

2. It was during this period that the last vestiges of subsistence agriculture disappeared from the Altar Valley, driven out by the depression, large operators such as the Gills and the Boices, and the Taylor Grazing Act, which ended homestead entries. Most smallholders in the Altar Valley were of Mexican descent, and information on them is scarce. The SCS reported in the late 1930s, "In the middle portion of the Altar, and on portions of the Pantano, there is [sic] a number of homesteaders, many of whom have attempted to make a living by dry farming, with but a few head of livestock at best to supplement their farming operations. These people have had considerable difficulty with ranchers over the distribution of land, according to some. At the present time their position is not well established in the country, and they, as a class, eke out a livelihood with considerable difficulty. As a rule, these homesteaders attempt a little general farming. In some

sections these homesteads have been largely abandoned and are now being utilized as ranch lands" (SCS 1936: 4–5).

3. "Projects and payments granted for each ranch in 1937 were as follows: development of springs and seeps, $50; earthen reservoirs, fifteen cents per cubic yard of excavation; wells (if water conveyed to a tank), $1 per linear foot; water spreading by permanent ditching, ten cents per linear foot; and range fences (at least three wires with good posts not more than twenty feet apart), thirty cents per rod; rodent control (destruction of at least ninety per cent of the rodents in an infected area), six to fifteen cents per acre; re-seeding, twenty cents per pound of seed sown; and fire guards (not less than four feet in width), three cents per linear foot" (Wagoner 1952: 62 n. 72).

4. The *Saturday Evening Post* reported that there were seventy-five watering sources in 1946 (Thruelsen 1946); Clayton Vincent, the Buenos Aires ranch manager, recalls sixty-four at the time the Gills sold the ranch in 1959 (personal communication, 25 July 1997).

5. R. H. Williams, "The Beef Outlook," in *Proceedings of the Arizona Cattle Growers' Association 8th Annual Meeting,* Douglas, Ariz., 4–6 January 1915 (Phoenix: Arizona Cattle Growers' Association, 1915), 21–25 (UASC archives).

6. This is not to say that producers were themselves unaware of concentrated power in the beef shipping and processing industries. To the contrary, complaints about high railroad rates, low prices, fraudulent or manipulative marketing, and packer concentration were periodically registered, especially at cattlemen's association conventions. These protests helped secure a variety of federal laws and regulations related to the transportation of live animals and the purity of processed beef (True 1937: 190–191). Typically, however, such regulations left corporate concentration little changed and actually helped meatpackers by certifying their products as safe, especially for export (Shumsky 1997).

7. The causes of vegetation change are still debated among ecologists: overgrazing, fire suppression (due to lack of grass cover as well as human efforts), seed dispersal by livestock, climate change, and erosion are all cited in various measures (Bahre 1991; Dobyns 1981; Hastings and Turner 1965; Humphrey 1958, 1962; McClaran and Van Devender 1995).

5 THE URBANIZATION OF RANCHING

1. Pamphlets cited in this section may be found at the University of Arizona Science Library among the publications of state agricultural experiment stations. I am indebted to Barbara Tellman for directing me to these mostly uncataloged materials.

2. Unless otherwise noted, information and quotations in this section come

from articles in the *Arizona Daily Star*, published on 2 January 1960, 18 February 1960, 10 March 1960, 26 April 1960, 19 May 1964, 23 May 1964, 30 June 1964, 1 July 1964, 2 July 1964, 4 July 1964, 9 July 1964, 29 July 1964, and 27 August 1964.

3. The difference in value between ranch lands and urban lands is much older, of course. C. B. Brown (1931) recorded that "Suburban Lands" were assessed at $57.38 per acre in 1931, "Irrigated Lands or Subject thereto" at $52.24 per acre, "Dry Farming Lands" at $9.75 per acre, and "Grazing Lands" at $2.57 per acre. But this is only a gap in values, not a gap in rent. Only later, when the prospect of developing ranch lands affected what people would pay for the land, did a rent gap emerge.

4. One solution to this problem, available only to ranchers with substantial personal fortunes, is to finance ranch debt themselves. If the ranch cannot pay off the loan, the owner-lender simply writes it off as a loss and credits the ranch books with a capital improvement. This has potential tax benefits and obviously allays any fears of foreclosure by creditors.

5. This change is conspicuous on the outskirts of Tucson, where earlier subdivisions tend to maintain 3.3-acre minimum lot sizes, whereas post-1963 developments cluster houses more densely around common areas, especially golf courses.

6. During an interview with a resident of a suburban development near the Buenos Aires Refuge, a rather mangy-looking coyote approached the large picture window in front of us. A Gambel's quail, perched on a stump just outside the window, scolded the coyote, which withdrew timidly. "I just love our wildlife," said the home owner.

6 RESTORING GRASS AND QUAIL

1. "The Victorio Company: Agricultural Real Estate Historic Investment Performance," January 1982. In 1969 Pruett and Wray purchased the Heady-Ashburn Ranch for just under $500,000; seven years later Victorio sold it for $1.76 million. A parcel at Palominas Farm increased in value from less than $90,000 to $200,000 in under four years. The Singing Valley Ranch, purchased in the summer of 1974 for $656,000, sold nine months later for more than $1 million. I am indebted to Wayne Pruett for sharing documents related to the Victorio Company and the Buenos Aires Ranch with me.

2. The gain of the Pozo Nuevo Ranch was offset by the sale of the Garcia Ranches on the southeast end of the Buenos Aires. Victorio reacquired this land, as well as several other parcels (including the La Osa Ranch), between 1972 and 1980.

3. Ratios varied somewhat, but the standard mixtures in 1979–82 were six parts blue panic, six parts Johnson, two parts sprangletop, and one part Lehmann in the bottomlands; two parts blue panic, two parts sprangletop, one part Lehmann, and one part Boer or plains in the uplands; two to three pounds seed per acre were sown in upland areas, and seven to eight pounds seed per acre were sown in the bottomlands. There is no documentation of the mixtures used from 1973 to 1978. Lehmann and Boer seeds were considerably more expensive than the others. The shoreline of Aguirre Lake received Jap millet and Hegari in equal measure. There is evidence that plains bristlegrass was seeded on one pasture (Trampa de Toro).

4. This figure should be taken as a conservative estimate. Some documents suggest a total as high as ninety thousand acres, but the majority of the difference probably represents repeat treatments.

5. An article in *Audubon* (Ricciuti 1979) amply documents the tortured bureaucratic politics surrounding Patuxent and its masked bobwhite work.

6. In the spring of 1982, eighty virile Texas bobwhites escaped from their pens. According to an internal memo, "most were recaptured" but "many remain in the wild" (memo from David Ellis to Steve Hoffman, 4 May 1982). If any of these birds survived to breed with released masked bobwhites, the subspecific genetic purity of the latter has been compromised, a grave violation of the Linnaean principles informing the Endangered Species Act and all FWS masked bobwhite work. It need hardly be added that this escape has been buried by the FWS, is mentioned nowhere in other official documents, and is unknown (or at least unspoken) among refuge employees.

7. According to David Ellis (personal communication, 23 February 2001), releases were discontinued according to plans that had been in place for some time. The timing was auspicious for the FWS: Had releases continued for several more years without further successes, the prospects for a refuge would have plummeted.

8. John Goodwin, personal communication, 4 May 1998.

9. David Brown has published the claim that after 1979, "uncontrolled grazing on the study pastures, combined with summer drought, resulted in sharply reduced populations" (1989: 137). The Buenos Aires ranch manager from 1980 to 1982 says that when he arrived, the release areas had been fenced off from cattle for several years and that no grazing of them occurred during his tenure; his successor, however, recalls that some "controlled grazing" did take place. Two other biologists involved in the efforts report that some grazing occurred throughout the period 1977–81 and that attempts to monitor grazing pressure statistically foundered because ranch managers failed to keep or provide the data.

10. John Goodwin, personal communication, 4 May 1998; and Steve Dobrott, personal communication, 16 April 1998.

11. See Goodwin (1982: 13–14) and documents of the Arizona State Land Department and the Buenos Aires National Wildlife Refuge pertaining to Pruett and Wray's range-improvement program. Birds released in 1975 entered pastures manipulated in 1974 and 1975; birds released in 1976 entered pastures manipulated in 1976–78; birds released in 1978 entered a pasture manipulated in 1973; and birds released in 1979 entered pastures manipulated in 1973–74 and 1978.

12. No adult birds were released after 1976. The following, however, were released: in 1976, 54 adults and 556 chicks; in 1977, 780 chicks and 68 juveniles; in 1978, 166 juveniles four to seven weeks of age; and in 1979, 535 chicks three to four weeks of age (Kuvlesky and Dobrott 1995: 24–25).

13. The revised Recovery Plan of 1984 actually called for the establishment of three separate and self-sustaining populations of masked bobwhite, and several other sites were discussed at the time. Until the population at the Buenos Aires Refuge is self-sustaining, however, the issue of additional refuges for the quail is dead. I have encountered no discussion of it among any FWS officials.

14. The Endangered Species Act does contain provisions for habitat preservation, but it is precisely on this point—where things become much more complex and constrained in actuality—that the FWS has found enforcement most difficult and public support least reliable. Indeed, the FWS publicly boasts that no private construction project has ever been scuttled on endangered species grounds.

7 PRODUCING A STATE OF NATURE

1. These include early drafts and summaries of the management plan, written by an FWS regional office employee (FWS 1986a, 1986b); comments on those documents by members of the Master Plan Team; revised versions sent to a large mailing list of interested parties; written comments received from agency personnel and members of the public; and finally, the draft management plan itself (FWS 1987). The materials reveal a gradual toning down of vocabulary and removal of more controversial statements. I have generally interpreted the earlier versions to reflect the original thinking and ambitions of the FWS with regard to the Buenos Aires Refuge, which were modified but not fundamentally changed by input from others.

2. Letter from State Land Commissioner Robert K. Lane to Regional Director Michael Spear, dated 6 October 1986; BLM/SLD "Land Exchange II" plan dated December 1987. I am indebted to Professor Jon Souder of Northern Arizona University for sharing with me files he obtained under the Freedom of Information Act pertaining to the refuge's State Trust leases.

3. "Refuge Planning Bulletin, Buenos Aires NWR, Edition 3," August 1986, p. 4.

4. The notes of one member of the Master Plan Team record that the FWS wanted "total control" of all of the State Trust lands as early as March 1986. These notes open as follows: "Schedule is as outlined so plan can be completed while [Governor] Babbitt is in office. Refuge planning has a priority over re-establishment of the masked bobwhite. Priority for FWS: *Public Use Policy*." The meeting took place at the Buenos Aires Refuge on 11 March 1986 and was led by the refuge manager.

5. "Refuge Planning Bulletin, Buenos Aires NWR, Edition 3," August 1986, p. 10.

6. Grazing, like other commercial uses of the refuge, may be permitted only if the refuge manager determines that it is "compatible with the primary purposes for which the area was established" (Curtin 1993: 34). Conducting a study would more or less obligate the refuge to make a formal compatibility determination based on the results. Instead, the refuge manager has simply refused to undertake any compatibility review, thereby closing the issue.

7. Biologist John Goodwin visited the ranch in May 1985 and afterward wrote to Regional Director Spear: "I have just returned from the Buenos Aires Ranch and was quite disturbed by much of what I saw. The overgrazing in many pastures is the worst I have seen in the past ten years" (internal memo). Goodwin estimated that there were only five hundred to six hundred acres of "suitable habitat" at that time, and he urged Spear not to release more birds than that area could support. "[R]elease of birds prior to the availability of suitable habitat *will fail*," he cautioned (emphasis in original).

8. Masked bobwhites are extremely skittish. When frightened they attempt to fly, sometimes slamming their heads into the roofs of their cages hard enough to hurt or even kill themselves. (Cages are extremely low roofed in an effort to minimize this trauma.) Disease is managed by sterilizing the facility and administering antibiotics in the birds' drinking water. Cannibalism is perhaps the trickiest problem. Refuge employees must decide whether the stress caused by debeaking and claw clipping is justified by the risk of mortality through pecking and clawing. Every brooder pen of like-aged birds has a "pecking order" from which escape is not an option.

9. Biologists both in and out of the FWS who have worked with the masked bobwhite describe the earlier technique as little more than a "hawk-feeding" exercise, in which the chicks were quickly abandoned by their surrogates and soon died. Obviously this characterization is a product of hindsight. Before giving up on Texas bobwhites, refuge biologists attempted to increase adoption rates by

implanting testosterone inhibitors. Research had found lower testosterone levels among "acceptors" than "rejectors"; acceptance rates increased from 46 to 58 percent (FWS 1989: 35).

10. In terms of numbers, "self-sustaining" is defined as an average of two hundred or more calling males over a five-year period, with no single year dropping below fifty calling males and no supplemental releases. This means an estimated five hundred birds, on average, surviving on their own.

11. The vast majority of paperwork is processed through computers; some of it may never assume printed form, although transactions involving money are always represented on paper. I employ the generic term "paperwork" to include computer work, as do refuge employees themselves.

12. It is not the mere existence of paperwork that sets the refuge apart. Paper representations played a role in the ranch as well, in land titles, leases, accounting, and so forth. Refuge management, however, involves a quantitative extension of paperwork to realms previously undocumented and to a level of detail not achieved on the ranch, resulting in a qualitative change in the social experience and meaning of production itself.

8 "WHERE WILDLIFE COMES NATURALLY!"

1. Eno and Di Silvestro (1985: 174). According to Amos Eno, who lobbied Congress on behalf of the National Audubon Society, the Arizona Cattle Growers' Association was less active in opposing the refuge than was its California equivalent (personal communication, 4 February 1999).

2. This decision may have been influenced by area residents' complaints that the refuge was bad for the local economy. By leading visitors to Sasabe, the route could be seen as a benefit for the town's one gas station/store. Asked generally about the refuge's impact on business, the store's owner made no mention of the auto route, however, and she did not report any positive effect on business from the refuge. The refuge is not far enough away from Tucson to require a second tank of gas, and most public use occurs at least seven miles away from Sasabe.

3. Former U.S. Secretary of the Interior and Arizona native Stewart Udall, who sided with the horses, countered: "I think if we could get the wildlife people to see it as a wildlife story and not a horse story—after all, horses are wildlife, a wonderful form of wildlife—then you've got something that's very compelling. . . . But government bureaucrats oftentimes don't like to be bothered by a thing that's a little complex" (*Arizona Republic*, 30 April 1990).

4. I generalize. Partly because of FWS policies, I did not conduct systematic surveys of visitors. During our time in Brown Canyon, the only group of visitors that was not wholly or disproportionately white was invited by my wife on the

basis of a personal relationship. They showed a lively interest in the canyon but did not consciously endeavor to see and identify wildlife. When the refuge hosted groups of FWS employees in the lodge, there were frequently minorities among them, mostly Native Americans and Mexican Americans. Among those who came to the canyon to see wildlife, however, I neither recall nor recorded a single nonwhite visitor. Judging from conversations, I surmised that the vast majority worked in white-collar professions or owned their own businesses.

9 COUNTERFEITING CONSERVATION

1. The Society for Environmental Truth is a small nonprofit organization formed in 1992 and based in Tucson. A promotional brochure for the society opens this way: "For the last twenty years we have been besieged and assaulted from all sides by extremist environmental organizations that tell us the world is doomed because of global warming, ozone depletion, destruction of natural resources, air pollution, water pollution and pesticides. We are told that there is a choice only between preservation and exploitation. But there *is* a third option and that is responsible management of natural resources. As a result of the scare tactics used by the extremists, our congress and legislators have passed restrictive environmental laws—many totally unnecessary—imposed costly regulations upon industry and consumers, and forced all of us to incur enormous costs which drain our monetary resources away from productive enterprises."

2. This was done very recently (see *Arizona Daily Star,* 20 July 1999), due largely to Brown's insistence. As Brown predicted in talking with me, however, the wild-caught birds were released along with the captive-bred ones, which Brown claims will make scientific evaluation of the results difficult or impossible and will result in the genetic-swamping of any wild birds that survive. Brown has long argued that releases of captive-bred birds should be halted for a number of years to determine if wild-caught birds can survive. "If they can't make it then, I'd say the bird can't be brought back in Arizona. We've created as good habitat as is reasonably possible, but the masked bobwhite is at the edge of its range, and erratic population dynamics may mean it can't succeed" (David Brown, personal communication, 20 May 1998).

3. Although it is clear that pronghorn once inhabited the valley, evidence of their abundance is scarce. I have been unable to locate any documentary evidence to support the refuge manager's claim, which recurs in the most recent draft management plan (FWS 2000: 57). There it is supported by testimony (attributed to homesteader Manuel King) that does not appear in the source cited for it.

References

ARCHIVAL SOURCES

Archival documents of particular significance are referenced in the text and listed in the bibliography.

AHS   Arizona Historical Society, Tucson
BANWR   Buenos Aires National Wildlife Refuge, Sasabe
CSWR   Center for Southwestern Research, Zimmerman Library, University of New Mexico, Albuquerque
GLO   General Land Office, microfiche records, housed at the offices of the U.S. Bureau of Land Management, Tucson and Phoenix
UASC   University of Arizona Library, Special Collections, Tucson

PUBLISHED SOURCES AND GOVERNMENT DOCUMENTS

Abruzzi, William S.
1995   The Social and Ecological Consequences of Early Cattle Ranching in the Little Colorado River Basin. *Human Ecology* 23(1): 75–98.

Agrawal, Arun
1998   *Greener Pastures: Politics, Markets, and Community among a Migrant Pastoral People.* Durham, N.C.: Duke University Press.

Aguirre, Yginio F.
1969   The Last of the Dons. *Journal of Arizona History* 10(4): 239–255.
1975   Echoes of the Conquistadores: Stock Raising in Spanish-Mexican Times. *Journal of Arizona History* 16(3): 267–286.
1986   Interview by David Siegel, U.S. Fish and Wildlife Service. Transcript, 20 November 1986. AHS and BANWR.

Allen, J. A.
1886a   The Masked Bobwhite *(Colinus ridgwayi)* in Arizona. *Auk* 3: 275–276.
1886b   The Masked Bob-white *(Colinus ridgwayi)* of Arizona, and Its Allies. *Bulletin of the American Museum of Natural History* 1(7): 273–290.
1887   A Further Note on *Colinus ridgwayi. Auk* 3: 483.

Alonso, Ana Maria
1994   The Politics of Space, Time, and Substance: State Formation, Nationalism, and Ethnicity. *Annual Review of Anthropology* 23: 379–405.

Anable, Michael E., Mitchel P. McClaran, and George B. Ruyle

1992   Spread of Introduced Lehmann Lovegrass *Eragrostis lehmanniana* Nees. in Southern Arizona, USA. *Biological Conservation* 61: 181–188.

Andrews, David A.

1937   *Ground Water in Avra-Altar Valley, Arizona*. U.S. Geological Survey Water-Supply Paper No. 796-E. Washington, D.C.: Government Printing Office.

Arizona State Land Department

1997   *Annual Report, 1996–97*. Phoenix: Interagency Printing Services.

Arrington, O. N.

1942   *A Survey of Possible Refuge Areas for Masked Bobwhite in the Arivaca and Altar Valley*. Special Report Project No. 11-D. Phoenix: Arizona Game and Fish Commission.

Atherton, Lewis

1961   *The Cattle Kings*. Bloomington: Indiana University Press.

Bahre, Conrad Joseph

1985   Wildfire in Southeastern Arizona between 1859 and 1890. *Desert Plants* 7(4): 190–194.

1991   *A Legacy of Change: Historic Human Impact on Vegetation in the Arizona Borderlands*. Tucson: University of Arizona Press.

Bahre, Conrad J., and Marlyn L. Shelton

1996   Rangeland Destruction: Cattle and Drought in Southeastern Arizona at the Turn of the Century. *Journal of the Southwest* 38(1): 1–22.

Barnes, Will Croft

1960   *Arizona Place Names*. 2d ed. Revised and enlarged by Byrd H. Granger. Tucson: University of Arizona Press.

Barthes, Roland

1957   *Mythologies*. Selected and translated by Annette Lavers. New York: Farrar, Straus and Giroux.

Belsky, A. J., A. Matzke, and S. Uselman

1999   Survey of Livestock Influences on Stream and Riparian Ecosystems in the Western United States. *Journal of Soil and Water Conservation* 54(1): 419–431.

Bentley, H. L.

1898   *Cattle Ranges of the Southwest: A History of the Exhaustion of the Pasturage and Suggestions for Its Restoration*. U.S. Department of Agriculture Farmer's Bulletin No. 72. Washington, D.C.: Government Printing Office.

1902   *Experiments in Range Improvement in Central Texas*. U.S. Depart-
            ment of Agriculture, Bureau of Plant Industry Bulletin No. 13.
            Washington, D.C.: Government Printing Office.

Bourdieu, Pierre

1977   *Outline of a Theory of Practice*. Translated by Richard Nice. Cam-
            bridge: Cambridge University Press.

1984   *Distinction: A Social Critique of the Judgement of Taste*. Translated
            by Richard Nice. Cambridge: Harvard University Press.

1998   *Practical Reason: On the Theory of Action*. Stanford: Stanford Uni-
            versity Press.

Bourne, Eulalia

1968   *Nine Months Is a Year at Baboquívari School*. Tucson: University of
            Arizona Press.

Bowden, Charles

1992   Baboquivari: A Place outside of Time. *Arizona Highways* (January):
            39–47.

Brenner, Neil

1998   Between Fixity and Motion: Accumulation, Territorial Organiza-
            tion, and the Historical Geography of Spatial Scales. *Environment
            and Planning D: Society and Space* 16: 459–481.

Brewster, William

1885   Additional Notes on Some Birds Collected in Arizona and the Ad-
            joining Province of Sonora, Mexico. *Auk* 2: 196–200.

Brown, C. B.

1928, 1929, Annual Reports of the Pima County Agricultural Extension Service.
1931   AHS.

Brown, David E.

1989   *Arizona Game Birds*. Tucson: University of Arizona Press and the
            Arizona Game and Fish Department.

Brown, David E., ed.

1982   Biotic Communities of the American Southwest—United States
            and Mexico. *Desert Plants* 4(1–4).

Brown, David E., and David H. Ellis

1977   *Status Summary and Recovery Plan for the Masked Bobwhite*. Albu-
            querque: U.S. Fish and Wildlife Service Office of Endangered Spe-
            cies, Region 2.

1984   *Masked Bobwhite Recovery Plan (1984 Revision)*. Revised by Ste-
            phen W. Hoffman. Albuquerque: U.S. Fish and Wildlife Service Of-
            fice of Endangered Species, Region 2.

Brown, Herbert

1885   Arizona Quail Notes. *Forest and Stream* 25(23): 445.

1900   The Conditions Governing Bird Life in Arizona. *Auk* 17: 31–34.

1904   Masked Bob-White *(Colinus ridgwayi)*. *Auk* 21(2): 209–213.

Browne, J. Ross

1871   *Adventures in the Apache Country*. New York: Harper & Bros.

Bryan, Kirk

1920   Origin of Rock Tanks and Charcos. *American Journal of Science* 50 (4th ser.): 188–206.

1922   Erosion and Sedimentation in the Papago Country, Arizona, with a Sketch of the Geology. In *Contributions to the Geography of the United States 1922*, pp. 19–85. U.S. Geological Survey Bulletin No. 730. Washington, D.C.: Government Printing Office.

1925   *The Papago Country, Arizona. A Geographic, Geologic, and Hydrologic Reconnaissance with a Guide to Watering Places*. U.S. Geological Survey Water-Supply Paper No. 499. Washington, D.C.: Government Printing Office.

1928   Change in Plant Associations by Change in Ground Water Level. *Ecology* 9(4): 474–478.

1929   Flood-Water Farming. *Geographical Review* 19: 444–456.

1940   Erosion in the Valleys of the Southwest. *New Mexico Quarterly* 10: 227–232.

Burton, Harley True

1928   *A History of the JA Ranch*. Austin, Tex.: Press of Von Boeckmann-Jones, Co.

Cable, D. R.

1965   Damage to Mesquite, Lehmann Lovegrass, and Black Grama by a Hot June Fire. *Journal of Range Management* 18: 326–329.

1971   Lehmann Lovegrass on the Santa Rita Experimental Range, 1937–1968. *Journal of Range Management* 24: 17–21.

Chadwick, Douglas H.

1996   Sanctuary: U.S. National Wildlife Refuges. *National Geographic* 190(4): 2–35.

Chapline, William R.

1944   The History of Western Range Research. *Agricultural History* 18: 127–143.

Cooke, Ronald U., and Richard W. Reeves

1976   *Arroyos and Environmental Change in the American South-West*. Oxford: Clarendon Press.

Coolidge, Dane

1939   *California Cowboys*. Tucson: University of Arizona Press.

Cooper, Toby

1978   Fish and Wildlife Director Downgrades Wildlife Help. *Defenders* 53(3): 174–175.

Coronil, Fernando

1997   *The Magical State: Nature, Money, and Modernity in Venezuela*. Chicago: University of Chicago Press.

Corrigan, Philip

1990   *Social Forms/Human Capacities: Essays in Authority and Difference*. New York: Routledge.

Corrigan, Philip, and Derek Sayer

1985   *The Great Arch: State Formation as Cultural Revolution*. Oxford: Basil Blackwell.

Coville, Frederick V.

1905   A Report on Systems of Leasing Large Areas of Grazing Land, Together with an Outline of a Proposed System for the Regulation of Grazing on the Public Lands of the United States. In *Grazing on the Public Lands: Extracts from the Report of the Public Lands Commission,* pp. 32–67. U.S. Department of Agriculture, Forest Service Bulletin No. 62. Washington, D.C.: Government Printing Office.

Cronon, William

1991   *Nature's Metropolis: Chicago and the Great West*. New York: Norton.

1996   The Trouble with Wilderness; or, Getting Back to the Wrong Nature. In *Uncommon Ground: Rethinking the Human Place in Nature,* edited by William Cronon, pp. 69–90. New York: Norton.

Curtin, Charles G.

1993   The Evolution of the U.S. National Wildlife Refuge System and the Doctrine of Compatibility. *Conservation Biology* 7(1): 29–38.

Curtiss, Charles F.

1898   *Some Essentials of Beef Production*. U.S. Department of Agriculture Farmers' Bulletin No. 71. Washington, D.C.: Government Printing Office.

Dart, Allen, James P. Holmlund, and Henry D. Wallace

1990   *Ancient Hohokam Communities in Southern Arizona: The Coyote Mountains Archaeological District in the Altar Valley*. Technical Report No. 90–93. Tucson: Center for Desert Archaeology.

Dary, David

1981   *Cowboy Culture: A Saga of Five Centuries*. New York: Knopf.

Davis, Goode P., Jr.

1970   Cowboys, Mountain Lions, and Politicians. *Defenders* 45(3): 292–
                296.

DeBuys, William, and Joan Meyers

1999   *Salt Dreams: Land and Water in Low-Down California.* Albuquerque:
                University of New Mexico Press.

Defenders of Wildlife

1978   Refuges for Wildlife, or Cows and Carnivals? *Defenders* 53(6): 307–
                311.

Dobyns, Henry F.

1981   *From Fire to Flood: Historic Human Destruction of Sonoran Desert
                Riverine Oases.* Anthropological Papers No. 20. Socorro, N.Mex.:
                Ballena Press.

Donahue, Debra L.

1999   *The Western Range Revisited: Removing Livestock from Public Lands
                to Conserve Native Biodiversity.* Norman: University of Oklahoma
                Press.

Duncklee, John

1994   *Good Years for the Buzzards.* Tucson: University of Arizona Press.

Dunlap, Thomas R.

1988   *Saving America's Wildlife.* Princeton: Princeton University Press.

Durkheim, Emile

1957   *Professional Ethics and Civic Morals.* Translated by Cornelia Brook-
                field. London: Routledge.

Dyksterhuis, E. J.

1949   Condition and Management of Range Land Based on Quantitative
                Ecology. *Journal of Range Management* 2(3): 104–115.

Ellickson, Robert C.

1991   *Order without Law: How Neighbors Settle Disputes.* Cambridge: Har-
                vard University Press.

Ellis, David H., Steven J. Dobrott, and John G. Goodwin, Jr.

1977   Reintroduction Techniques for Masked Bobwhites. In *Endangered
                Birds: Management Techniques for Preserving Threatened Species,*
                edited by Stanley A. Temple, pp. 345–353. Madison: University of
                Wisconsin Press.

Ellis, D. H., and J. A. Serafin

1977   A Research Program for the Endangered Masked Bobwhite. *Journal
                of the World Pheasant Association* 2: 16–33.

Emory, William H.

1857   *Report on the United States and Mexican Boundary Survey.* House of Representatives Ex. Doc. No. 135. Washington, D.C.: Cornelius Wendell, Printer.

Eno, Amos S., and Roger L. Di Silvestro

1985   *Audubon Wildlife Report 1985.* New York: National Audubon Society.

Ferguson, Denzel, and Nancy Ferguson

1984   Abused Oasis: Oregon's Malheur. *Defenders* 59(5): 20–29.

Ferguson, James

1994   *The Anti-politics Machine: Development, Depoliticization, and Bureaucratic Power in Lesotho.* Minneapolis: University of Minnesota Press.

Fischer, Hank

1977   Refuges in Trouble, Our Survey Discloses. *Defenders* 52(4): 281–283.

Fleischner, Thomas L.

1994   Ecological Costs of Livestock Grazing in Western North America. *Conservation Biology* 8(3): 629–644.

Fleischner, T. L., D. E. Brown, A. Y. Cooperrider, W. B. Kessler, and E. L. Painter

1994   Society for Conservation Biology Position Statement: Livestock Grazing on Public Lands in the United States of America. *Society for Conservation Biology Newsletter* 1(4): 2–3.

Fowler, John M., and James R. Gray

1988   Rangeland Economics in the Arid West. In *Rangelands,* edited by Bruce A. Buchanan, pp. 67–89. Albuquerque: University of New Mexico Press.

Frink, Maurice, W. Turrentine Jackson, and Agnes Wright Spring

1956   *When Grass Was King.* Boulder: University of Colorado Press.

Gallizioli, Steve

1976   Deadly Overgrazing. *Defenders* 51(3): 161–163.

Gallizioli, Steve, Seymour Levy, and Jim Levy

1967   Can the Masked Bobwhite Be Saved from Extinction? *Audubon Field Notes* 21(5): 571–575.

Galloway, B. T.

1913   The Bureau of Plant Industry, Its Functions and Efficiency. In *Miscellaneous Papers,* pp. 3–12. U.S. Department of Agriculture, Bureau of Plant Industry Circular No. 117. Washington, D.C.: Government Printing Office.

Geiger, E. L., and G. R. McPherson

2000   Changes in Semi-desert Grassland Plant Communities Ten Years Following Cessation of Grazing and Reintroduction of Fire. Paper presented at the 85th annual meeting of the Ecological Society of America, Snowbird, Utah, 6–10 August.

Goldman, O. B., and Robert H. Burns

1923   Preliminary Report, La Osa Ranch. Unpublished ms. Kinney files, AHS.

Goodwin, John G., Jr.

1982   Habitat Needs of Masked Bobwhite in Arizona. University of Arizona contract report to U.S. Fish and Wildlife Service. BANWR.

Graham, Richard

1960   The Investment Boom in British-Texan Cattle Companies, 1880–1885. *Business History Review* 34: 421–445.

Granger, Byrd H.

1983   *Arizona's Names: X Marks the Place.* Tucson: Falconer Publishing Co.

Gray, Gary G.

1993   *Wildlife and People: The Human Dimensions of Wildlife Ecology.* Urbana: University of Illinois Press.

Gressley, Gene M.

1959   Teschemacher and deBillier Cattle Company: A Study of Eastern Capital on the Frontier. *Business History Review* 33: 121–137.

1966   *Bankers and Cattlemen.* New York: Knopf.

Griffiths, David

1901   *Range Improvement in Arizona.* U.S. Department of Agriculture, Bureau of Plant Industry Bulletin No. 4. Washington, D.C.: Government Printing Office.

1904   *Range Investigations in Arizona.* U.S. Department of Agriculture, Bureau of Plant Industry Bulletin No. 67. Washington, D.C.: Government Printing Office.

1907   *The Reseeding of Depleted Range and Native Pastures.* U.S. Department of Agriculture, Bureau of Plant Industry Bulletin No. 117. Washington, D.C.: Government Printing Office.

1910   *A Protected Stock Range in Arizona.* U.S. Department of Agriculture, Bureau of Plant Industry Bulletin No. 177. Washington, D.C.: Government Printing Office.

Guthery, Fred S., Nina M. King, Kenneth R. Nolte, William P. Kuvlesky, Jr., Stephen DeStefano, Sally A. Gall, and Nova J. Silvy.

2000 Comparative Habitat Ecology of Texas and Masked Bobwhites. *Journal of Wildlife Management* 64(2): 407–420.

Haley, J. Evetts

1953 *The XIT Ranch of Texas and the Early Days of the Llano Estacado.* Norman: University of Oklahoma Press.

Harvey, David

1982 *The Limits to Capital.* Oxford: Basil Blackwell.

1985a The Geopolitics of Capitalism. In *Social Relations and Spatial Structures,* edited by Derek Gregory and John Urry, pp. 128–163. London: Macmillan.

1985b *The Urbanization of Capital: Studies in the History and Theory of Capitalist Urbanization.* Baltimore: Johns Hopkins University Press.

1996 *Justice, Nature, and the Geography of Difference.* Oxford: Blackwell.

2000 *Spaces of Hope.* Berkeley: University of California Press.

Haskett, Bert

1935 Early History of the Cattle Industry in Arizona. *Arizona Historical Review* 6: 3–42.

Hastings, James Rodney

1959 Vegetation Change and Arroyo Cutting in Southeastern Arizona. *Journal of the Arizona Academy of Science* 1(2): 60–67.

Hastings, James Rodney, and Raymond M. Turner

1965 *The Changing Mile: An Ecological Study of Vegetation Change with Time in the Lower Mile of an Arid and Semiarid Region.* Tucson: University of Arizona Press.

Hays, Samuel P.

1959 *Conservation and the Gospel of Efficiency: The Progressive Conservation Movement, 1890–1920.* Cambridge: Harvard University Press.

Heffernan, William D., and Douglas H. Constance

1994 Transnational Corporations and the Globalization of the Food System. In *From Columbus to ConAgra: The Globalization of Agriculture and Food,* edited by Alessandro Bonanno et al., pp. 29–51. Lawrence: University Press of Kansas.

Henderson, George L.

1999 *California and the Fictions of Capital.* New York: Oxford University Press.

Hendrickson, Dean A., and W. L. Minckley

1984   Ciénegas—Vanishing Climax Communities of the American Southwest. *Desert Plants* 6(3): 131–175.

Herbel, Carlton H., and Robert P. Gibbens

1996   *Post-drought Vegetation Dynamics on Arid Rangelands of Southern New Mexico.* Agricultural Experiment Station Bulletin No. 776. Las Cruces: New Mexico State University.

Hirt, Paul

1989   The Transformation of a Landscape: Culture and Ecology in Southeastern Arizona. *Environmental Review* 13(3–4): 167–189.

1994   *A Conspiracy of Optimism: Management of the National Forests since World War II.* Lincoln: University of Nebraska Press.

Holling, C. S., and Gary K. Meffe

1996   Command and Control and the Pathology of Natural Resource Management. *Conservation Biology* 10: 328–337.

Hudson, Ray

2001   *Producing Places.* New York: Guilford Press.

Humphrey, Robert R.

1958   *The Desert Grassland: A History of Vegetational Change and an Analysis of Causes.* Tucson: University of Arizona Press.

1962   *Range Ecology.* New York: Ronald Press Co.

1987   *Ninety Years and 535 Miles: Vegetation Changes along the Mexican Border.* Albuquerque: University of New Mexico Press.

Hymel, Mona L.

1998   Tax Policy and the Passive Loss Rules: Is Anybody Listening? *Arizona Law Review* 40: 615.

Ingold, Tim

1980   *Hunters, Pastoralists, and Ranchers: Reindeer Economies and Their Transformations.* Cambridge Studies in Social Anthropology No. 28. Cambridge: Cambridge University Press.

Jessop, Bob

1990   *State Theory: Putting Capitalist States in Their Place.* University Park: Pennsylvania State University Press.

Johnson, Stephen Aubrey

1973a   Arizona—The Bounty State. *Defenders* 48(5): 535–536.

1973b   Arrogance Revisited: The Mohave County Rancher and the Mountain Lion. *Defenders* 48(4): 413–415.

1974   Bring Them in Regardless of How. *Defenders* 49(2): 119–121.

1978a   Ranching: High Cost of an American Myth. *Defenders* 53(6): 324–327.

1978b   Western Ranching: The Cost of the Myth. *Defenders* 54(2): 119–120.

1985   Just *Whose* Home Is the Range? *Defenders* 60(4): 14–17.

Johnson, Terry B., and Stephen W. Hoffman

n.d.   *The Masked Bobwhite: A Critical Decision*. Phoenix: Arizona Game and Fish Department in conjunction with U.S. Fish and Wildlife Service.

Jordan, Terry G.

1993   *North American Cattle-Ranching Frontiers: Origins, Diffusion, and Differentiation*. Albuquerque: University of New Mexico Press.

Joseph, Gilbert M., and Daniel Nugent, eds.

1994   *Everyday Forms of State Formation: Revolution and the Negotiation of Rule in Modern Mexico*. Durham, N.C.: Duke University Press.

Karl, Thomas R., and Richard W. Knight

1985   *Atlas of Monthly Palmer Moisture Anomaly Indices (1931–1984) for the Contiguous United States*. Asheville, N.C.: National Climatic Data Center.

Kerlinger, Paul

1993   Birding Economics and Birder Demographics Studies as Conservation Tools. In *Status and Management of Neotropical Migratory Birds,* edited by Deborah M. Finch and Peter W. Stangel, pp. 32–38. U.S. Department of Agriculture, Forest Service, General Technical Report No. RM-229. Fort Collins, Colo.: Rocky Mountain Forest and Range Experiment Station.

King, Nina Monique

1998   Habitat Use by Endangered Masked Bobwhites and Other Quail on the Buenos Aires National Wildlife Refuge, Arizona. Master's thesis, University of Arizona.

Kinney, William C.

1996   Conditions of Rangelands before 1905. In *Sierra Nevada Ecosystem Project: Final Report to Congress,* vol. 2, *Assessments and Scientific Basis for Management Options,* pp. 31–45. Davis: University of California, Centers for Water and Wildland Resources.

Kloppenburg, Jack Ralph Jr.

1988   *First the Seed: The Political Economy of Plant Biotechnology, 1492–2000*. Cambridge: Cambridge University Press.

Knipe, Theodore

n.d.   A Preliminary Wildlife Habitat Study of the Buenas *[sic]* Aires Ranch. BANWR.

Knobloch, Frieda

1996   *The Culture of Wilderness: Agriculture as Colonization in the American West.* Chapel Hill: University of North Carolina Press.

Kuvlesky, William P., Jr., and Steve J. Dobrott

1995   Masked Bobwhite Recovery Plan, Second Revision. Albuquerque: U.S. Fish and Wildlife Service Region 2. BANWR.

Larmer, Forrest M.

1926   *Financing the Livestock Industry.* New York: Macmillan.

Lawrence, Elizabeth Atwood

1982   *Rodeo: An Anthropologist Looks at the Wild and the Tame.* Chicago: University of Chicago Press.

Leavengood, Betty

1993   History of the Buenos Aires National Wildlife Refuge. Unpublished ms. BANWR.

Lefebvre, Henri

1977   *De l'etat 3: Le mode de production étatique.* Paris: Union Générale d'Éditions.

1991   *The Production of Space.* Translated by Donald Nicholson-Smith. Oxford: Basil Blackwell.

Lewis, David Rich

1994   *Neither Wolf nor Dog: American Indians, Environment, and Agrarian Change.* New York: Oxford University Press.

Liffmann, Robin H., Lynn Huntsinger, and Larry C. Forero

2000   To Ranch or Not to Ranch: Home on the Urban Range? *Journal of Range Management* 53: 362–370.

Ligon, J. Stokley

1927   *Wild Life of New Mexico: Its Conservation and Management. Being a Report on the Game Survey of the State, 1926 and 1927.* Santa Fe: State Game Commission.

1946   *Upland Game Bird Restoration through Trapping and Transplanting.* University of New Mexico Publications in Biology No. 2. Albuquerque: University of New Mexico Press.

1952   The Vanishing Masked Bobwhite. *Condor* 54: 48–50.

1961   *New Mexico Birds and Where to Find Them.* Albuquerque: University of New Mexico Press.

Lloyd, W. A.

1924   An Extension Program in Range Livestock, Dairying, and Human
            Nutrition for the Western States. U.S. Department of Agriculture
            Circular No. 308. Washington, D.C.: Government Printing Office.

MacCannell, Dean

1976   The Tourist: A New Theory of the Leisure Class. New York: Schocken
            Books. 2d ed., 1989.

Martin, William E., and Gene L. Jefferies

1966   Relating Ranch Prices and Grazing Permit Values to Ranch Produc-
            tivity. Journal of Farm Economics 48(2): 233–242.

Mattison, Ray H.

1946   Early Spanish and Mexican Settlements in Arizona. New Mexico
            Historical Review 21(4): 273–327.

McClaran, Mitchel P., and Thomas R. Van Devender, eds.

1995   The Desert Grassland. Tucson: University of Arizona Press.

McGee, W. J.

1897   Sheetflood Erosion. Bulletin of the Geological Society of America 8:
            87–112.

McKenzie, Evan

1994   Privatopia: Homeowners' Associations and the Rise of Private Gov-
            ernment. New Haven: Yale University Press.

McPherson, Guy R.

1995   The Role of Fire in the Desert Grasslands. In The Desert Grassland,
            edited by Mitchel P. McClaran and Thomas R. Van Devender, pp.
            130–151. Tucson: University of Arizona Press.

McPherson, Guy R., and Jake F. Weltzin

2000   Disturbance and Climate Change in United States/Mexico Border-
            land Plant Communities: A State-of-the-Knowledge Review. U.S. De-
            partment of Agriculture, Forest Service, General Technical Report
            No. RMRS-GTR-50. Fort Collins, Colo.: Rocky Mountain Research
            Station.

Mills, G. Scott

1984   Survey of the Buenos Aires Ranch for Masked Bobwhite and Other
            Birds with an Assessment of Habitat Suitability—August 1984.
            Unpublished ms. prepared for the U.S. Fish and Wildlife Service.
            BANWR.

Mitchell, Timothy

1988   Colonising Egypt. Berkeley: University of California Press.

1991 The Limits of the State: Beyond Statist Approaches and Their Critics. *American Political Science Review* 85(1): 77–96.

Morrisey, Richard J.

1950 The Early Range Cattle Industry in Arizona. *Agricultural History* 24: 151–156.

Munk, Joseph A.

1905 *Arizona Sketches*. New York: Grafton Press.

Murray, Justin

1973 Rotten Ranching. *Defenders* 48(5): 532.

Nelson, Richard

1997 *Heart and Blood: Living with Deer in America*. New York: Knopf.

Nelson, Robert H.

1995 *Public Lands and Private Rights: The Failure of Scientific Management*. Lanham, Md.: Rowman and Littlefield Publishers.

Osgood, Ernest Staples

1929 *The Day of the Cattleman*. Minneapolis: University of Minnesota Press. Reprint, Chicago: University of Chicago Press, n.d.

Ough, William D., and James C. deVos

1982 Masked Bobwhite Investigations on the Buenos Aires Ranch in August 1982. Contract report from the Arizona Game and Fish Department, Special Services Division, Field Contracts Section to the U.S. Fish and Wildlife Service, Endangered Species Office. Contract No. 14-16-0002-82-216. BANWR.

Phillips, Allan R., Joe Marshall, and Gale Monson

1964 *The Birds of Arizona*. Tucson: University of Arizona Press.

Pickrell, Charles U.

1925 *The Range Bull*. University of Arizona College of Agriculture, Extension Service Circular No. 51. Tucson: University of Arizona.

Pieper, Rex D., Jerry C. Anway, Mark A. Ellstrom, Carlton H. Herbel, Robert L. Packard, Stuart L. Pimm, Ralph J. Raitt, Eugene E. Staffeldt, and J. Gordon Watts

1983 *Structure and Function of North American Desert Grassland Ecosystems*. Agricultural Experiment Station Special Report No. 39. Las Cruces: New Mexico State University.

Pima County Board of Supervisors

2000 *Our Common Ground: Ranch Lands in Pima County. Sonoran Desert Conservation Plan Preliminary Ranch Element*. Tucson: Pima County Board of Supervisors.

Pisani, Donald

1989   The Irrigation District and the Federal Relationship. In *The Twentieth Century West: Historical Interpretations,* edited by Gerald Nash and Richard Etulain, pp. 257–292. Albuquerque: University of New Mexico Press.

Pitkin, Hanna Fenichel

1984   *Fortune Is a Woman: Gender and Politics in the Thought of Niccolò Machiavelli.* Berkeley: University of California Press.

Pomeroy, Earl

1957   *In Search of the Golden West: The Tourist in Western America.* New York: Knopf.

Postone, Moishe

1993   *Time, Labor, and Social Domination: A Reinterpretation of Marx's Critical Theory.* Cambridge: Cambridge University Press.

Potter, Albert F.

1905   Questions regarding the Public Grazing Lands of the Western United States. In *Grazing on the Public Lands: Extracts from the Report of the Public Lands Commission,* pp. 11–31. U.S. Department of Agriculture, Forest Service Bulletin No. 62. Washington, D.C.: Government Printing Office.

Powell, John Wesley

1879   *Report on the Lands of the Arid Region of the United States, with a More Detailed Account of the Lands of Utah.* 2d ed. Washington, D.C.: Government Printing Office.

Public Lands Commission

1905   Extract from the Second Partial Report of the Public Lands Commission. In *Grazing on the Public Lands,* pp. 7–10. U.S. Department of Agriculture, Forest Service Bulletin No. 62. Washington D.C.: Government Printing Office.

Pumpelly, Raphael

1870   *Across America and Asia. Notes of a Five Years' Journey around the World and of Residence in Arizona, Japan, and China.* New York: Leypoldt and Holt.

Pyne, Stephen J.

1982   *Fire in America: A Cultural History of Wildland and Rural Fire.* Princeton: Princeton University Press.

Rasmussen, Wayne D.

1989   *Taking the University to the People: Seventy-five Years of Cooperative Extension.* Ames: Iowa State University Press.

Reisner, Marc

1986   *Cadillac Desert: The American West and Its Disappearing Water*. New York: Penguin. Revised ed., 1993.

Remley, David

1993   *Bell Ranch: Cattle Ranching in the Southwest, 1824–1947*. Albuquerque: University of New Mexico Press.

Reynolds, H. G., and F. H. Tschirley

1963   *Mesquite Control on Southwestern Rangeland*. U.S. Department of Agriculture Leaflet No. 421. First issued in 1957 as Leaflet No. 234. Washington, D.C.: Government Printing Office.

Ricciuti, Edward R.

1979   Deathwatch at Patuxent. *Audubon* (January): 82–92.

Ritvo, Harriet

1987   *The Animal Estate: The English and Other Creatures in the Victorian Age*. Cambridge: Harvard University Press.

Robinett, D.

n.d.   The History, Soil, and Plant Resources of the Altar Valley. U.S. Department of Agriculture, Soil Conservation Service, unpublished ms. BANWR.

1994   Fire Effects on Southeastern Arizona Plains Grasslands. *Rangelands* 16(4): 143–148.

Roosevelt, Theodore

1888   *Ranch Life and the Hunting-Trail*. Reprint, Lincoln: University of Nebraska Press, 1983.

Rose, Carol M.

1994   *Property and Persuasion: Essays on the History, Theory, and Rhetoric of Ownership*. Boulder, Colo.: Westview Press.

Roskruge, George

1885–86   Field notes of surveys of the Altar Valley, conducted under the authority of the General Land Office. GLO.

Roundy, Bruce A., and Sharon H. Biedenbender

1995   Revegetation in the Desert Grassland. In *The Desert Grassland*, edited by Mitchel P. McClaran and Thomas R. Van Devender, pp. 265–304. Tucson: University of Arizona Press.

Rowe, Helen Ivy, E. T. Bartlett, and Louise E. Swanson, Jr.

2001   Ranching Motivations in Two Colorado Counties. *Journal of Range Management* 54: 314–321.

Runte, Alfred
1979   *National Parks: The American Experience.* Lincoln: University of
            Nebraska Press.

Russell, Stephen M.
1986   What Price Preservation? In *Masked Bobwhite Biology and Conserva-*
            tion, edited by Mark R. Stromberg, Terry B. Johnson, and Stephen
            W. Hoffman, pp. 18–21. Proceedings of a symposium, University of
            Arizona, Tucson, 6 December 1984. Phoenix: Arizona Game and
            Fish Department and the National Audubon Society.

Sampson, Arthur W.
1919   *Plant Succession in Relation to Range Management.* U.S. Depart-
            ment of Agriculture Bulletin No. 791. Washington, D.C.: Govern-
            ment Printing Office.

Sayre, Nathan F.
1999   The Cattle Boom in Southern Arizona: Towards a Critical Political
            Ecology. *Journal of the Southwest* 41(2): 239–271.
2001   *The New Ranch Handbook: A Guide to Restoring Western Rangelands.*
            Santa Fe: Quivira Coalition.

Schellie, Don
1970   *The Tucson Citizen: A Century of Arizona Journalism.* Tucson: Citi-
            zen Publishing Co.

Scott, James C.
1998   *Seeing Like a State: How Certain Schemes to Improve the Human*
            Condition Have Failed. New Haven: Yale University Press.

Scott, Martha
1978   Defender's Survey: Refuge Policies Belie the Name. *Defenders* 53(1):
            68–69.

Sewell, William H., Jr.
1992   A Theory of Structure: Duality, Agency, and Transformation. *Amer-*
            ican Journal of Sociology 98(1): 1–29.

Shepard, Paul
1991   *Man in the Landscape: A Historic View of the Esthetics of Nature.*
            College Station: Texas A&M University Press.

Sheridan, Thomas E.
1995   *Arizona: A History.* Tucson: University of Arizona Press.
2001   Cows, Condos, and the Contested Commons: The Political Ecology
            of Ranching on the Arizona-Sonora Borderlands. *Human Organiza-*
            tion 60(2): 141–152.

Shumsky, Michael
1997   Government Regulatory Policy and the Struggle for Meat Inspec-
            tion, 1879–1906. Electronic document. At www.fas.harvard.edu/
            šhumsky/inspection.html.
Skaggs, Jimmy M.
1986   *Prime Cut: Livestock Raising and Meatpacking in the United States,
            1607–1983*. College Station: Texas A&M University Press.
Slatta, Richard W.
1990   *Cowboys of the Americas*. New Haven: Yale University Press.
Smith, Allen E.
1982   It's Time to Stop Wildlife Refuge Giveaways. *Defenders* 57(4): 44.
Smith, Arthur H., and William E. Martin
1972   Socioeconomic Behavior of Cattle Ranchers, with Implications for
            Rural Community Development in the West. *American Journal of
            Agricultural Economics* 54(2): 217–225.
Smith, Henry Nash
1978   *Virgin Land: The American West as Symbol and Myth*. Reprint of
            1950 ed. Cambridge: Harvard University Press.
Smith, Jared G.
1895   *A Note on Experimental Grass Gardens*. U.S. Department of Agri-
            culture Circular No. 1. Washington, D.C.: Government Printing
            Office.
1899   *Grazing Problems in the Southwest and How to Meet Them*. U.S.
            Department of Agriculture, Division of Agrostology Bulletin No.
            16. Washington, D.C.: Government Printing Office.
Smith, Neil
1984   *Uneven Development: Nature, Capital, and the Production of Space*.
            Oxford: Basil Blackwell.
1996   *The New Urban Frontier: Gentrification and the Revanchist City*. Lon-
            don: Routledge.
Sonnichsen, C. L.
1950   *Cowboys and Cattle Kings: Life on the Range Today*. Norman: Univer-
            sity of Oklahoma Press.
Souder, Jon A., and Sally K. Fairfax
1996   *State Trust Lands: History, Management, and Sustainable Use*. Law-
            rence: University Press of Kansas.
Southern Arizona Cattlemen's Protective Association (SACPA)
1996   *The Buenos Aires National Wildlife Refuge, 1985–1995: The Re-
            introduction Program of the Masked Bobwhite Quail*. Tucson: South-

ern Arizona Cattlemen's Protective Association and the Society for Environmental Truth.

Starrs, Paul F.

1998   *Let the Cowboy Ride: Cattle Ranching in the American West.* Baltimore: Johns Hopkins University Press.

Stegner, Wallace

1954   *Beyond the Hundredth Meridian: John Wesley Powell and the Second Opening of the West.* Penguin ed., 1992. New York: Houghton Mifflin.

Stephens, F.

1885   Notes of an Ornithological Trip in Arizona and Sonora. *Auk* 2(3): 225–231.

Stocking, George

1995   Wild Kingdom. *Arizona Adventure* (9 February): 9–11.

Stoddard, Herbert L.

1931   *The Bobwhite Quail: Its Habits, Preservation, and Increase.* New York: Scribner's.

Stromberg, Mark R., Terry B. Johnson, and Stephen W. Hoffman, eds.

1986   *Masked Bobwhite Biology and Conservation.* Proceedings of a symposium, University of Arizona, Tucson, 6 December 1984. Phoenix: Arizona Game and Fish Department and the National Audubon Society.

Swetnam, Thomas W., and Julio L. Betancourt

1998   Mesoscale Disturbance and Ecological Response to Decadal Climatic Variability in the American Southwest. *Journal of Climate* 11: 3128–3147.

Tellman, Barbara, Richard Yarde, and Mary G. Wallace

1997   *Arizona's Changing Rivers: How People Have Affected the Rivers.* Water Resources Research Center Issue Paper No. 19. Tucson: University of Arizona College of Agriculture.

Thompson, E. P.

1967   Time, Work-Discipline, and Industrial Capitalism. *Past and Present* 38: 56–97.

Thornber, J. J.

1910   *The Grazing Ranges of Arizona.* Agricultural Experiment Station Bulletin No. 65. Tucson: University of Arizona.

Thruelsen, Richard

1946   Cow Brutes Are Like People. *Saturday Evening Post* (17 August): 26, 78–82.

Thu, Kendall M., and E. Paul Durrenberger, eds.

1998   *Pigs, Profits, and Rural Communities*. Albany: State University of New York Press.

Tomlinson, Roy

1972a   Current Status of the Endangered Masked Bobwhite Quail. In *Transactions of the Thirty-Seventh North American Wildlife and Natural Resources Conference*, pp. 294–311. 12–15 March 1972. Washington, D.C.: Wildlife Management Institute.

1972b   *Review of Literature on the Endangered Masked Bobwhite*. U.S. Fish and Wildlife Service Resource Publication No. 108. Washington, D.C.: Government Printing Office.

Toumey, J. W.

1891   *Notes on Some of the Range Grasses of Arizona and Overstocking the Range*. Agricultural Experiment Station Bulletin No. 2. Tucson: *Arizona Citizen*.

Trow-Smith, Robert

1959   *A History of British Livestock Husbandry, 1700–1900*. London: Routledge and Kegan Paul.

True, Alfred Charles

1937   *A History of Agricultural Experimentation and Research in the United States, 1607–1925*. Miscellaneous Publication No. 251. Washington, D.C.: U.S. Department of Agriculture.

University of Arizona Agricultural Extension Service (UAAES)

1959   *Your Range: Its Management*. Compiled by Robert R. Humphrey. Special Report No. 2. Tucson: University of Arizona College of Agriculture.

Urban, Sharon

1982   History of the Altar Valley. In *Background Studies for Environmental Assessment, Southern Arizona Auxiliary Airfield*, vol. 2. Draft report prepared for the Air Directorate, National Guard Bureau by the Benham Group, Oklahoma City, Okla. Unpublished ms. BANWR and Office of the State Archaeologist, Arizona State Museum, Tucson.

U.S. Department of Agriculture (USDA)

1898   *Yearbook of the Department of Agriculture 1897*. Washington, D.C.: Government Printing Office.

1900   *Yearbook of the Department of Agriculture 1899*. Washington, D.C.: Government Printing Office.

U.S. Department of the Interior, Census Office

1883   *Report on the Production of Agriculture as Returned at the Tenth Census (June 1 1880)*. Washington, D.C.: Government Printing Office.

U.S. Department of Transportation

1978   *Manual on Uniform Traffic Control Devices for Streets and Highways.* Washington, D.C.: Federal Highway Administration.

U.S. Fish and Wildlife Service (FWS)

1984a   Buenos Aires Ranch Proposed Protection of Masked Bobwhite Habitat, Pima County, Arizona. Public hearing, University of Arizona, Tucson, Ariz., 24 July 1984. Transcribed by Bouley, Schlesinger, DiCurti & Schippers. BANWR.

1984b   Environmental Assessment regarding Buenos Aires Ranch. Public hearing, Sasabe, Ariz., 25 July 1984. Transcribed by Bouley, Schlesinger, DiCurti & Schippers. BANWR.

1985   *Department of the Interior Final Environmental Assessment: Protection of Masked Bobwhite Habitat, Pima County, Arizona.* Albuquerque: U.S. Fish and Wildlife Service Region 2. BANWR.

1986a   Long-Range Management Strategy. Unpublished ms. BANWR.

1986b   Prospectus of the Master Plan for Buenos Aires NWR. Unpublished ms. BANWR.

1987   Master Plan for Buenos Aires National Wildlife Refuge, Pima County, Arizona. U.S. Department of the Interior Draft Environmental Assessment. Albuquerque: U.S. Fish and Wildlife Service Region 2. BANWR.

1988   Annual Narrative Report, Calendar Year 1988. BANWR.

1989   Annual Narrative Report, Calendar Year 1989. BANWR.

1990   Annual Narrative Report, Calendar Year 1990. BANWR.

1993   Annual Narrative Report, Calendar Year 1993. BANWR.

1995   Annual Narrative Report, Calendar Year 1995. BANWR.

1996   Grazing Issues at Buenos Aires NWR. BANWR.

1997   *Banking on Nature: The Economic Benefits to Local Communities of National Wildlife Refuge Visitation.* Prepared by Andrew Laughland and James Caudill. Arlington, Va.: U.S. Fish and Wildlife Service Division of Refuges.

1998   Packet of materials distributed to participants in the Buenos Aires National Wildlife Refuge Comprehensive Conservation Plan Workshops in Tucson, Phoenix, and Arivaca in March. BANWR.

2000   *Buenos Aires National Wildlife Refuge Draft Comprehensive Conser-vation Plan and Environmental Assessment*. Albuquerque: U.S. Fish and Wildlife Service Region 2.

U.S. Secretary of the Interior

1881–1911   *Report of the Governor of Arizona to the Secretary of the Interior*. Washington, D.C.: Government Printing Office.

U.S. Soil Conservation Service (SCS)

1936   Ultra-extensive Report on Santa Cruz Watershed. CSWR.

1938   Santa Cruz Watershed: Preliminary Examination Report. CSWR.

1992   *Brawley Wash Natural Resource Restoration Plan, Pima County, Ari-zona*. Phoenix: U.S. Department of Agriculture.

Wagoner, J. J.

1952   *History of the Cattle Industry in Southern Arizona, 1540–1940*. Social Science Bulletin No. 20. Tucson: University of Arizona.

Walton, John

1992   *Western Times and Water Wars: State, Culture, and Rebellion in California*. Berkeley: University of California Press.

Wasser, C. H.

1977   Early Development of Technical Range Management, ca. 1895–1945. *Agricultural History* 51(1): 63–77.

Webb, Walter Prescott

1931   *The Great Plains*. Reprint, 1981. Lincoln: University of Nebraska Press.

Westoby, Mark, Brian Walker, and Imanuel Noy-Meir

1989   Opportunistic Management for Rangelands Not at Equilibrium. *Journal of Range Management* 42(4): 266–274.

Whitaker, Bob

1986   Return of the Bandit Quail. *Arizona Highways* 62(6): 32–37.

White, Richard

1983   *The Roots of Dependency: Subsistence, Environment, and Social Change among the Choctaws, Pawnees, and Navajos*. Lincoln: University of Nebraska Press.

1991   *It's Your Misfortune and None of My Own: A New History of the American West*. Norman: University of Oklahoma Press.

Wilbur-Cruce, Eva Antonia

1987   *A Beautiful, Cruel Country*. Tucson: University of Arizona Press.

Williams, Raymond

1980   *Problems in Materialism and Culture: Selected Essays*. London: Verso.

Wooton, E. O.

1916   *Carrying Capacity of Grazing Ranges in Southern Arizona*. U.S. Department of Agriculture Bulletin No. 367. Washington, D.C.: Government Printing Office.

Workman, J. P., and J. M. Fowler

1986   Optimum Stocking Rate: Biology vs. Economics. In *Rangelands: A Resource Under Siege*, edited by P. J. Joss et al., pp. 101–102. Proceedings of the Second International Rangeland Congress, Adelaide, South Australia. Cambridge: Cambridge University Press.

Worster, Donald

1985   *Rivers of Empire: Water, Aridity, and the Growth of the American West*. New York: Oxford University Press.

1992   *Under Western Skies: Nature and History in the American West*. New York: Oxford University Press.

1997   The Wilderness of History. *Wild Earth* 7(3): 9–13.

Wuerthner, George

1994   Subdivisions versus Agriculture. *Conservation Biology* 8(3): 905–908.

index

classifying, 8–10; as commodity, 223–24; distribution of, xxxvi, 10–12, 19–21, 235nn. 1–3, 236nn. 9, 11, 12; as endangered species, 3, 22, 150–51; fire and, 166–67; grazing and, 143–44; management of, 144–45, 161–64, 226–27, 242nn. 6–8; public and, 187–88; reintroduction of, xlv, xlvii–xlviii, 16–18, 126, 127–29, 135–42, 146–47, 148, 149–50, 243nn. 11–13, 244nn. 9, 10, 246n. 2; research on, 24–25; SACPA on, 222–23; values of, xxv, xxvi, 25–26

bobwhite, northern, 7–8

bobwhite, Texas, 139, 161, 242n. 6, 244n. 9

Bourdieu, Pierre: on capital, xxiv–xxvi

Brenner, Pablo, 145, 146, 161

Brewster, William, 9

Brown, C. B., 84, 85–87, 88, 220

Brown, David, 137, 138(fig.), 139, 226, 242n. 9

Brown, Herbert: and bobwhite, 5–8, 10–13, 25, 142, 235n. 2

Brown, Rollin, 201

Brown Canyon: public use of, 200, 201–5, 211–13, 245n. 4; signage at, 172–74

Browne, J. Ross, xliii–xliv, 35

Bryan, Kirk, 44, 71–72

Buenos Aires National Wildlife Refuge (BANWR), xv, xxxvi–xxxvii, 151, 153, 217–18; and Brown Canyon, 202–4, 245n. 4; as bureaucracy, 175–80, 245nn. 11, 12; debates over, 147–48, 221–27, 245nn. 1, 2; establishment of, 129, 140, 141, 152; fire use on, 165–67; grazing on,

242n. 9, 244nn. 6, 7; land for, 191–92; management plan for, 154–60, 186–87, 243n. 1, 244n. 4; masked bobwhites on, 161–64; "nature" on, xix, 167–69, 230–31; pronghorns on, 229–30, 246n. 3; public and, xlviii–xlix, 186–91, 196–201; signage on, 169–75, 194–96(fig.); symbolism of, 181–83, 245n. 3

Buenos Aires Ranch (Ayres Ranch), xv, xxxvi, 30, 34–35, 37, 38, 49, 56, 68, 77, 105, 238n. 1; drought and, 99–100; FWS and, 146–48; management of, xlvi–xlvii, 42–43, 50–51, 58, 75, 82, 83–84, 88–94, 104, 119–20, 126, 130–32, 148, 220–21, 227, 243n. 3; masked bobwhites on, 127–29, 135–42, 143–44, 145, 149–50, 243nn. 11–13; mesquites on, 103, 132–33; pasture on, 133–35

built environment, xxiii–xxiv, 119

bureaucracies: Buenos Aires Refuge as, 169–80, 245n. 11; land management, 58–59

Bureau of Animal Industry, 96

Bureau of Plant Industry, 59, 63

burning: prescribed, xv–xvi, 165–67. *See also* fire

cactus, Pima pineapple, 168–69

Camou, José, 68, 69, 72, 74

Campos, Alfredo, 113

capital, xxiii; bureaucratic, 154, 195, 215–17; and cattle, 31–32, 43, 52–54, 130–31, 238n. 11; ecology and, 124–25; La Osa's, 68–69; ranching and, 116–18; types of, xxv–xxvi; Victorio Company, 129–30

(USDA), xxi, 63, 96; agrostology and, 59–61

U.S. Department of the Interior, 56, 58, 184

U.S. Fish and Wildlife Service (FWS), xv, xxi, xxiv, xxxvi, xlvii; and Buenos Aires, 129, 145, 146–48, 152, 154–69, 182, 188–90, 202–3, 217–18, 222–23, 226–27; as bureaucracy, 175–80, 215–16; and masked bobwhite, 3, 18, 22–25, 127–28, 136–42, 242nn.6, 7; and public use, 184–86

U.S. Forest Service (USFS), 58, 65, 78, 102, 136

U.S. Geological Survey (USGS), xx, 71

U.S. Soil Conservation Service (SCS), xxi, xlvi, 83, 89, 238n. 9

University of Arizona Agricultural Experiment Station, 100–101

University of Arizona Agricultural Extension Service, 97–98

urbanization, 107–8, 114, 115, 231; land values and, 123–24; lawns and, 109–11

value(s): ecological, 123–24; of endangered species, 150–51; property, 122–23, 158–59; symbolic, xxv, xxvi, 24–25, 230–31; of wildlife, 186–87, 204–15

Victorio Company, xlvii–xlviii, 219, 241nn. 1, 2; restoration by, 126–27, 128–30, 132–35, 144–45, 148–49, 227

Warren Ranch, 35, 38, 48. *See also* Anvil Ranch

Watchable Wildlife campaign, 194–95, 196(fig.), 199

water, xxvii, 54, 124; in Altar Valley, xxxv, xxxvi, xl, xlii, xliii, 30, 32–33, 35, 49, 71–72, 93, 237nn. 1, 2, 5, 240n. 4; at Buenos Aires, 132, 152–53, 189–90; development of, xlv–xlvi, 73–74, 104, 240n. 3

watersheds, 85–87, 89–90, 93–94

wells, xlii, xliv, xlvi, 32–33, 35, 37–38, 44, 132

Wilbur-Cruce, Eva Antonia, 192

Wilbur-Cruce property, 193–94

wildlife, xxxviii, 25, 121, 137n. 13: in Brown Canyon, 203–4; on Buenos Aires, 93, 157–58; as capital, 215–17; counting, 199–201; public and, 21, 182, 183, 186–88, 189, 194–95, 196, 205–6, 245n. 4

wildlife refuges, 184–86

World War I, 58, 65, 84

Wray, Peter, 126, 129

Wright, Willard, 47, 48

About the Author

Nathan F. Sayre is assistant professor of geography at the University of California, Berkeley. After college he worked for the Arizona Conservation Corps on the Buenos Aires National Wildlife Refuge and other public lands, constructing fences to control cows, trails to control hikers, and barricades to control off-highway vehicles. He is the author of *The New Ranch Handbook: A Guide to Restoring Western Rangelands* (Quivira Coalition, 2001) and *Working Wilderness: The Malpai Borderlands Group and the Future of the Western Range* (Rio Nuevo Press, 2005). He lives in Berkeley with his wife, Lucia, and their two children, Henry and Lila.